"What a fun read! A
with so much of wl
recommend Miss Ke.
air talent."

—Amy Vanderoef, Host of *Good Morning Texas*

"Hilary Kennedy has provided a remarkable gift for anyone interested in the inner workings of the entertainment industry. Her diary is a refreshingly candid, remarkably insightful journal, intimately conveying the daily challenges, obstacles, joy and satisfaction that go into building a career as a successful TV personality. Not only does Hilary take you behind the scenes for a first-hand glimpse of 'the business,' but she provides countless helpful hints and useful strategies that have helped her break in – and stay in – the most competitive of fields. I recommend Hilary Kennedy's book to anyone interested in establishing a career in the entertainment industry – or establishing a career period."

—Richard J. Allen, former head writer, *Days of Our Lives*, two-time Emmy winner, *As the World Turns*

"Hilary Kennedy's diary is a great tool for all aspiring TV hosts to read, study and learn from. It has all the dos and don'ts that Hilary has uncovered from her real life experience. The path to becoming a recognized TV personality is a long and hard one, but as you will see after reading her diary, Hilary's strength, perseverance and faith have enabled her to get this far which is farther than the vast majority of those who set foot on this career path."

—Larry Namer, founder of E! Entertainment Television Network

"Hilary Kennedy's Career Diary offers much in regards to the realities of working in television, including in-depth accounts of interviewing and scheduling conflicts. Her account is bright and informative, and an asset to anyone venturing into the realm of hosting."

—Pat Summerall, former NFL star and broadcaster

"In this case, you can judge a book by its cover. Hilary's journey towards success is not reality with a clever plot to hook you in; it's an honest look inside the struggles and triumphs of a successful television host. Anyone who dreams her same dream can learn a lot by reading this story of dedication to her passion. This is Hilary's reality."

—Cheryl Elliott, *Beauty and the Geek* (Season One)

"A must read for anyone considering becoming on-air talent. Hilary shows that being a television host is more than being able to read a teleprompter. A true professional is dedicated to the network, the show, as well as the crew and Hilary gives perfect examples of how to accomplish just that. Entertaining as much as it is informative, a true to life look into the world of a television host."

—Dave Murden, executive producer/director, ColdCut Films

CAREER DIARY™

OF A

TV SHOW HOST

Thirty days behind the scenes with a professional.

GARDNER'S CAREER DIARIES™

HILARY KENNEDY

GARTH GARDNER COMPANY

GGC publishing

Washington DC, USA · London, UK

Editorial inquiries concerning this book should be mailed to:
The Editor, Garth Gardner Company, 5107 13th Street N.W., Washington DC 20011 or emailed to: info@ggcinc.com. http//:www.gogardner.com

Copyright © 2007, GGC, Inc. All rights reserved.

No part of this book may be reproduced, stored in a retrieval system, or transmitted in any form or by any other means—electronic, mechanical, photocopying, recording, or otherwise —except for citations of data for scholarly or reference purposes, with full acknowledgment of title, edition, and publisher, and written notification to GGC/Publishing prior to such use. GGC/Publishing is a department of Garth Gardner Company, Inc. and Gardner's Career Diaries is a registered trademark of Garth Gardner Company, Inc.

ISBN 1-58965-044-1

Library of Congress Cataloging-in-Publication Data

Kennedy, Hilary.
  Career diary of a TV show host : thirty days behind the scenes with a professional / Hilary Kennedy.
       p. cm. -- (Gardner's Career Diaries)

ISBN 1-5895-044-1

1. Interviewing on television--Vocational guidance. 2. Television--Production and direction--Vocational guidance. 3. Kennedy, Hilary. 4. Television personalities--United States--Biography. 5. Career I. Title.
   PN1992.8.I68K46 2007
   791.4502'8092--dc22
   [B]
                                    2007030898

Printed in Canada

TABLE OF CONTENTS

7 **Biography**

9 **Current Position & Responsibilities**

13 **Resume**

17 Day 1

23 Day 2

27 Day 3

31 Day 4

35 Day 5

39 Day 6

43 Day 7

47 Day 8

51 Day 9

57 Day 10

61 Day 11

67 Day 12

73 Day 13

79 Day 14

83 Day 15

87 Day 16

93 Day 17

97 Day 18

101 Day 19

105 Day 20

109 Day 21

115 Day 22

119 Day 23

125 Day 24

129 Day 25

133 Day 26

137 Day 27

141 Day 28

145 Day 29

149 Day 30

155 **Glossary**

ACKNOWLEDGMENTS

First, I would like to thank the Lord for blessing me in every area of my life. I would like to acknowledge my friends and family whom assisted me with this diary, offering their creative input as well as their beautiful faces. I appreciate my camera operators, Chris Lunardon and Robert Morris, helping me both in front of and behind the camera on this project. I owe Scott Robertson a great deal for affording me my first experience as a host, and for his wonderful guidance. To my grandparents, Julie, Donny, Bill, and Joshua-thank you for cheering me on. Lastly, special thanks to my mom, Suzanne, who has been my most loyal supporter and avid follower of the show. Mom, I am truly your biggest fan.

BIOGRAPHY

My name is Hilary Kennedy, and I'm a full-time host for a television show in a number-six market. I earned my Bachelor's degree in Radio–Television–Film (with a minor in Theatre) from Texas Christian University. Upon graduation I was signed by a well-known talent agency, having first been declared a finalist after entering a national network casting call. Reading a career diary written by an actress taught me a great deal about the entertainment business, and also gave me ideas for submitting letters to casting agents and marketing myself as a talent. I began auditioning regularly for commercials and industrials (business-oriented instructional films) and took supplemental acting classes in my spare time. I worked for several marketing companies to support myself financially, where I promoted high-end products at events throughout the city. That gave me the scheduling flexibility to audition for acting jobs and also served to hone my people skills.

I was fairly lucky my first year in the business, landing a national commercial within six months. I kept up regular contact with my agent, making sure my file was full of up-to-date headshots and resumes, and I did my best to forge friendly relationships with other actors. After shooting several local and regional commercials, I began sending out submission tapes to agencies on the West

Coast, where the greatest number of acting opportunities exist. After a visit out there to check on things firsthand, plus interviews with various agents and managers, I decided my career would be better served in smaller markets where the competition was less cutthroat. Although I considered myself a serious actor, the role of host was just another way to gain on-camera experience and pay the bills. My break came with the Dallas Stars, a National Hockey League team, when they hired me as their first female entertainment host. I worked to become comfortable while performing live in front of as many as 18,000 fans in the Dallas arena. I also shot promotional commercials for the team and interviewed several Hall of Fame players, which garnered me significant coverage in several national magazines. When a dispute between league management and the players' union caused an entire season to go up in smoke, my job disappeared.

I worked that following summer on several independent films and a few commercials, but my previous experience as a host compelled me to find employment in that field. I assembled a demo reel, pulling together highlights of my work with the Dallas Stars, and sent out copies to casting directors and network TV officials. I ended up guest hosting an automotive segment for a news network and also landed a role as the anchor for a local news-and-entertainment station. Jobs were steady, though unpredictable, and I spent the next two years traveling to Los Angeles, New York and Las Vegas to remain employed.

Once traveling became too time-consuming and exhausting, I decided it was time for a permanent on-camera hosting gig. After learning that the current host was leaving, I sent a submission letter and a demo reel to a well-established home-and-lifestyle show I'd viewed many times. Following two on-camera auditions and an interview with their vice president, I was offered a full-time, salaried position as the host.

CURRENT POSITION AND RESPONSIBILITIES

As the host of a weekly television show, my duties go far beyond what people see on their TV screens. Our corporate office has established certain standards of quality, and I collaborate with my managing producer to ensure that each show meets them. I also fulfill the role of associate producer, working with each assignment's director of photography (DP) and film editor to guarantee that our shows tell a story in a highly creative, entertaining and informative manner.

I'm responsible for the layout of the show, deciding what we cover, where we shoot, and which special features to emphasize. There is a lot of paperwork to fill out as well, such as waivers for all the people who appear on camera with me, and various internal documents that help our production crew keep everything focused.

Creativity is an important part of this job. I write my own introductions and decide on each show's theme, making sure the script stays true to it. I decide whom to interview and then build a series of questions that will best inform the viewers during each respective program. Together with the DP we select specific places to set up our shots, keeping in mind that we want to show off the location in the best possible light and also make the show entertaining for our viewers.

At each location I'm in charge of deciding the order in which my subjects are interviewed. I also do my best to help everyone relax, since many of my subjects are unaccustomed to being on television. Because the host becomes the de facto public face of a show, I'm always conscious of behaving properly when in the public eye. I make countless personal appearances to promote the show.

In post-production I review all the tapes and check for errors of fact or continuity, and I'm also responsible for ensuring that the order of the show makes sense to those watching it. I record voice-overs for use during the show, as well as promos that we use to publicize the program.

I'm also accountable to our clients, since they're the ones who pay the bills. I work hard to create unique relationships with each one, and I'm in regular contact with their marketing or promotional people to make sure each segment we shoot shows them in the best possible light. We do our best to accommodate their special needs, even if that means changing our shooting schedule.

I've been very proactive in signing up show sponsors, especially those who have traded wardrobe, salon styling and tanning facilities for promotional considerations. I make sure my viewers know who these sponsors are, and I encourage their patronage as a way to offset expenditures that would normally be coming out of my pocket or the show's budget.

Finally, since I'm an on-camera personality, there are a whole host (no pun intended) of responsibilities that are part of this position, including hair and makeup skills, the ability to read off a TelePrompTer, and memorize a script, as well as the talent to recognize when it's time to do just one more take to satisfy everyone, including myself.

Most of all, a host should have the people skills to handle and resolve all problems in a calm and rational manner, even under circumstances when everyone else may be in full panic mode.

RESUME

FILM & TV
Starring and co-starring roles:
- *Studio 13* (National Lampoon Network)
- *Laws of Deception*
- *The Evil Behind You*
- *Diwali*
- *Lovesick*
- *Believe*

HOSTING ROLES:
- *New Year's Nation*
- *Guitars and Cadillacs*
- *Philip Morris Nationwide Concert Simulcast*
- *The Local Attractions Channel*
- *Hot on Homes*

THEATRE
Starring roles:
- *A Beautiful Murder*
- *Decision at Hempstead*
- *Elizabeth the Queen*

- *I Hate Hamlet*
- *Impromptu*
- *Murder for the Holidays*
- *The Murder Game*
- *Clubhouse Crew Children's Sketches*

COMMERCIAL, INDUSTRIAL, RADIO
Principal or host:

- Alcon
- American Airlines
- Bankers' Compliance Training
- CBS/Grammy Awards
- Dallas Stars Hockey Team
- Germantown Aesthetics
- Hospice Health Care
- J.C. Penney's (national spot)
- Kiwi Carpet Services
- Lucky's Auto Repair
- Paws and Tails
- Pier One
- SiteVid
- Sonic Restaurants
- Spike Global Grill

- WingStop (national spot with Troy Aikman)

Featured:
- 7-Eleven (national spot)
- Denver Water Company
- Dick's Sporting Goods
- Garden Ridge
- Johnny Carino's Restaurant (national spot)

EDUCATION & TRAINING

B.A. in Radio–Television–Film–Theatre, Texas Christian University

- On-camera commercial training, KD Studio.
- Private instruction in acting, cold reading, improvisation, on-camera film study and monologue preparation.

SPECIAL SKILLS

Athletics:
- Cycling, dance, horseback riding, kickboxing, mountain climbing, Pilates, tennis, yoga, whitewater rafting.

Acting:
- Action & fight sequences, blue-screen, dialects, ear-prompt, TelePrompTerDiary.

THESE ARE SAMPLES OF MY PRESS KIT MATERIALS—THE DEMO REEL DVD AND CASE.

Day 1 JANUARY 15

PREDICTIONS

- Attend a Realtors' event to film various b-roll, interviews and stand-ups, with a five o'clock deadline
- Set up an interview for an upcoming Valentine's Day show

DIARY

The first day of the workweek is intentionally laid back, designed for preparing the shooting schedule for the rest of the week. I host a television home-and-lifestyle show in a Top Ten market, which means we average 200,000 viewers a week. Landing this job took a lot of hard work and strategic planning, but the rewards have been worth it. I'd always been fascinated with the life of a TV host—traveling all over the place, interviewing celebrities, and getting paid to be me! Now that I have my very own weekly TV show, I can reflect on which elements worked to get me here, and which ones did not.

I report to the production office early every Monday morning and work a standard 40-hour week. The person who schedules our filming tries to keep Mondays and Fridays clear so we can use that time to write our segments (dialogue), preview the shows that are nearing broadcast, and take care of other host-type duties. We try to schedule no more than one shoot a day, but oftentimes that turns into two, simply because we have so much ground to cover. Our crew travels to different home communities and lifestyle spots around the city to interview the "on-sites," the folks managing each of these respective venues or homes. I interview them on camera, separately film my introduction to each segment—that's known as a "stand up"—and our director of photography (DP) arranges to shoot some b-roll (background detail) that rounds out each show.

I feel a bit relieved after looking over this week's schedule. Wednesday is the only day I'll have two shoots in a row. However, once a month I'm asked to create a theme show, and this month's theme is Valentine's Day. Theme shows require a lot of planning and creativity, and they're designed to hook viewers into watching. My theme show for Christmas included Santa Claus as my co-host, and we filmed it in a gorgeous historic mansion. As great as it was, it put pressure on me to be just as creative and memorable with our Valentine's Day show. Last week I rang up a local winery known for its romantic ambiance, and they jumped at the chance to be featured. Now I need to compose some entertaining material to use as an introduction to each segment, making sure to stick to the wine theme.

I begin with the following: "Welcome to the special Valentine's Day edition of the show. I'm your host, Hilary Kennedy. It's common knowledge that love and wine go hand in hand, so this week we are coming to you from Delaney Vineyards. They specialize in perfect, romantic blends to share for a special occasion like this. While I learn more about the wine-making process, why don't you take a look at our first exciting community...?"

After writing eight additional stand-ups, I call to check on the status of my press kit. The role of a TV host is as much about press coverage as it is about being talented, smart and well prepared. In fact, I landed this job because of my high profile among members of the press. After being hired as the first female entertainment host for a major sports franchise, I sent out a press release to half a dozen entertainment and sports magazines—all national in coverage—as well as local media sources like our daily newspaper. Three national magazines and two Internet publications reported my story, which boosted my visibility for other, more lucrative jobs down the road. Even though I host my own show and have good viewer numbers, press coverage is still vital. My new press kit is an audio-visual presentation on DVD. The jewel case has my pertinent business information printed on the back cover, the same way a movie DVD includes a plot synopsis back there. The DVD itself contains highlights from my current show plus a number of important personal TV moments. In the world of television this is known as a "demo," since it *demonstrates* my on-camera personality.

I call to check on the production status and discover they're having trouble with the format I used to submit my material. Oh, great—another problem! It's been four months since I started working on this project. I hired a graphic designer to create the cover, the text and the artwork on the DVD itself. Once these issues are resolved and the printing is complete, I've already identified a number of media contacts to receive it.

Writing stand-ups and dealing with the press kit takes up most of my morning. After lunch I visit my clothing sponsor. In return for a billboard credit at the end of each show and a voice-over that announces the company's name and where they're located, they provide me with five outfits a week. This arrangement allows me to wear the latest fashions on camera without blowing my entire salary on new clothes every week. At the store one of the managers walks through the racks with me, where we pull out seven or eight outfits. I model them for fit and camera appeal, always mindful that some clothes look terrific in real life but not necessarily on TV.

My next stop is home, but only briefly. I try to attend weightlifting or yoga classes on Monday nights. Physical fitness is clearly important for anyone who appears on television, but working out seems to do wonders for my self-esteem as well. I use yoga to relax and de-stress, while the weightlifting tones my muscles and helps keep my weight consistent. I decided some time ago to eat a huge breakfast and lunch, but only a snack for dinner. That controls my food intake so I'll burn most of my calories while I'm working.

In the evening I often watch national sports and entertainment shows to see what other hosts are doing. It's sort of like "TV viewing as job research." I pay close attention to hand movements and other physical gestures to judge what looks natural on camera. I also analyze how these people deliver their lines, whether as a memorized script, read directly off a TelePrompTer, or delivered off the cuff.

LESSONS/PROBLEMS

I need to start planning for theme shows a little earlier so I can find ways to make them more interesting and creative. I also need to start thinking of how I want to package my demo reel and where it should be sent first.

THIS IS THE WIRELESS MIC, WHICH CLIPS TO THE BACK OF THE PANTS AND CAN BE HIDDEN BY RUNNING IT UP THROUGH A SHIRT OR BLOUSE.

Day 2 JANUARY 16

PREDICTIONS

- Shoot several interviews, plus some generic stand-ups
- Have dinner with my friend Scott

DIARY

My shoot this morning consists of three interviews and one stand-up, and all of them should be relatively easy. Last night I managed to properly upload my press kit information, and it's being printed today. Also, a commercial I shot several months ago has just started its run on national television. A number of my friends and family members called to say they saw it on prime time TV last night, and I'm looking forward to watching it myself.

I pack up my laptop and we're off to the shoot. Morris, my director of photography today, drives the van that's filled with our lighting and camera equipment. When we arrive at the location, everything is in order and the interviewees are prepared. By that I mean they're dressed appropriately, rehearsed as to what we'll discuss, and fairly relaxed about being on camera. These folks know exactly what to expect: brief, concise questions from me that should be answered in complete sentences while looking directly at me, not the camera. Morris sets up our lights and camera equipment while I get my interview questions and other paperwork in order. One important aspect of being a television host is to have signed waivers from everyone you plan to include in the show. I always have my subjects sign a waiver before we sit down in front of the camera. Once that's finished, I fill out the slate with the name of the person I'm interviewing plus their job title, if they have one. Then I add the show's ID number. I also write down the tape number so my editors know which program we're shooting. Like most shows, we tape from three to four weeks in advance of the actual airdate, so it can be confusing if we don't keep perfect track of all these details. To make the scripting of a segment easier on everyone—the editors and myself included—I use a maximum-capacity, large-screen portable multimedia player (PMP), which supports all standard video formats in full DVD resolution. I record just the responses to my questions while making note of the time code for each sound bite. After our interview session is over, I log all the usable sound bites onto a spreadsheet—including the time code, so I'll know exactly how long each piece runs—and then number these segments in the order in which I'll want them to appear. Some hosts have their associate or managing producers do this, but I prefer the freedom of deciding on my own what ends up where.

After my interviews are finished, I log everything we've shot onto my laptop. Meanwhile, Morris takes interior and exterior shots of the home and the surrounding community. Our primary focus is to film cool, exciting new things in home design and construction, plus the hot places to go in the city. Part of my job as host is to introduce these segments in a fun, natural and inviting manner. I find that fluid hand movements, a low and pleasing voice, and a kind smile are the best factors to employ while delivering my lines. When using a hand-held microphone, it's better to gesture in small ways with the free hand to prevent looking like a statue. As a television host, I'm often responsible for putting on my own microphone, especially if it's wireless, so I make sure to mount it somewhere that won't distract the viewer. I usually clip the transmitter to the back of my trousers or dress, while the wire that connects the unit to the microphone itself can be run up underneath my clothing for better concealment. I like to attach that end to my collar, while men usually clip it to their tie.

Later I make a call regarding the upcoming weekend. I've been approached to emcee an event for a nationally known charitable organization. This is one additional aspect of being a television host. Various organizations often ask you to donate your time to help others or raise awareness for a specific cause. The contact person there tells me I'll be making announcements, awarding prizes, interviewing celebrities, and playing a trivia game on stage with the patrons. They've also scheduled a celebrity chef cook-off for which I'll provide some sort of play-by-play commentary. This is the only aspect of the event that concerns me. I'm not much of a cook and I don't know anything about the participating chefs, so I'll need to do some research. The worst thing for a host is to arrive unprepared at an event, especially one with a live audience. I'll do some investigating tomorrow in between filming segments.

As the workday comes to a close, I pack up and head to my friend Scott's house for "movie night," which is a pleasant distraction from my busy day. I make a mental note to obtain some footage of work I did back when I hosted a national simulcast on New Year's Eve. That will add nicely to my press kit, yet another e-mail I'll need to send out as soon as I get home tonight.

LESSONS/PROBLEMS

I always enjoy hearing from other people who have seen my work and liked it, so the news about my commercial was appreciated. I know I need to prepare more for this weekend so the charity will be pleased with my performance and I'm either invited back next year, or other people will see my work and request me for their events. Tomorrow's goal will be to find a way to make this weekend's event a success—for the charity and for me.

A GRIP HELPS ARRANGE THE LIGHTS FOR THE SHOOT.

Day 3 JANUARY 17

PREDICTIONS

- With today's poor weather it's possible that some of my shoots will be rescheduled, which could push them into the weekend

DIARY

A phone call from my managing producer—the person to whom I report—wakes me up. We had snow overnight, so our morning shoot has been cancelled and I don't need to show up at the office until noon. I elect to sleep in...and why not? With my busy schedule I rarely have that luxury, so I decide to take advantage of the situation.

After lunch I head to the office. For some reason our Internet connection is down, which prevents me from researching the chefs for my weekend event. I'll have to squeeze that in later. Today we have both a director of photography (DP) and a grip assigned to the shoot, which will make things go more smoothly. While the DP is setting up a shot, the grip will look after the lighting and other details.

Because of the icy roads it takes us longer than usual to reach our location, but we still make it with time to spare. The house where we're shooting is in the million-dollar range. These homes generate considerable viewer interest, since they're the type of place many people dream about living in some day. While my crew sets up the equipment, I return a missed phone call. The producer of a show I used to host several times a year has asked me to come back and shoot again, if I have the time. I call back and tell them I'd be delighted to do it. Not only is this extra cash in my pocket, but also additional television exposure. One of the best things a host can do is enjoy maximum public visibility, so I make it a rule to never turn down a television appearance unless it's in direct conflict with my existing contract. One of the perks of my current job is the lack of an exclusivity agreement. In other words, I'm allowed to host other events and participate in commercials, so long as there isn't any obvious conflict of interest.

My interviews go fairly smoothly, although the first woman employs a few curse words when she messes up her answers. However, both of the women I interview seem willing to give me what I need for this segment, which are concise answers. Run-on sentences and the use of "um" and "you know" are distracting; although that's the way most people speak in informal conversation. Today my subjects are cognizant of their speech and manage to eliminate most of these unnecessary words.

On our way back to the office, a disturbing phone call comes in from someone at corporate headquarters. I'm told that 30 percent of the company was laid off this morning. Ouch! This is a scary situation, because one of the first things you learn as a host is that television gigs can be fleeting. Often there is no rhyme or reason to why a show is cancelled, or why someone else replaces the on-screen talent. I've been the host of some very promising ventures, only to have these programs cancelled a few months later or taken in a new direction that involves a different personality as host. After returning to work we're given the list of layoffs. Thankfully my name is not on it, but I almost feel guilty for having held onto my job.

At my desk I send out some e-mails to tell friends and family the startling news. The weather is still pretty nasty outside, so I decide to leave early and head home without stopping off at the gym. There I make a list of the new footage I'll need to add to my demo reel, contacting the production companies electronically that have the required material on hand. I know they'll gladly send it to me. I also do some online research regarding other TV hosts, both in my market and beyond. It's always a good idea to keep up with the competition to see how they've carved out a niche for themselves. And after today's news, it can't hurt to stay posted regarding what's going on in some of the larger markets.

One important thing to remember in these kinds of situations is that I wouldn't be the host of my own show if my hard work hadn't paid off. Hosting may not save the planet, but I believe people watch my show because it's an escape from the disarray of the outside world. It's light and fun, and it gives people 30 minutes to relax and forget about the dog that needs to be let out, or the roof that needs repair, or the kids who have after-school commitments from now until forever.

LESSONS/PROBLEMS
In the entertainment world, things can change in the blink of an eye. It's a lesson I know but still have difficulty with. It's important to have a number of solid contacts handy in case something happens, such as today's layoffs. On another note, I never got around to doing my cooking research, and I definitely don't want to show up unprepared this weekend.

PARTY SET-UP FOR A GRAND OPENING THEMED SHOOT.

Day 4 JANUARY 18

PREDICTIONS

- Conduct research on my upcoming charity event and figure out what they'll want in an emcee
- See how yesterday's layoffs have affected office morale

DIARY

My entire trip to work involves sitting in the worst possible traffic. My mood hardly improves when I reach the office. The heavy cloud that loomed over everyone yesterday still seems to be here. There's hardly any laughing or joking around in either the corporate or the production offices—nothing but somberness everywhere. I sit down at my computer and check my e-mail, look over today's shooting schedule, and then sit in silence for a while. I find it difficult to get excited about my show when so many negative things are happening behind the scenes.

At 10 o'clock I drive over to a community grand opening with one of my coworkers. We're planning to film the ribbon-cutting ceremony and then interview people at the party that follows. The event is well planned and everyone is excited to be part of the festivities, so I experience a momentary lift in my mood. Standing in front of the camera certainly helps, especially as I work hard to prepare my interviewees for their five seconds of fame. Before long we're finished with the assignment and it's time to head to lunch.

As the two of us are finishing our meal, we see The Truck. To advertise our show, our production company has hired a giant billboard truck to drive around town. The first time I saw it I felt proud but also a bit embarrassed, since it includes a huge photo of my face on it. My coworker and I flag it down at the curb and pose in front of it for a few pictures, which is fun.

Afterwards we head back to the office where I receive a call from the coordinator for this weekend's charity affair. The NHL All-Star Game and the Dennis Leary Firefighters Foundation are sponsoring it, a collaboration that has made national headlines. The list of events has grown to include a 5k race for both runners and in-line skaters, a street hockey game between celebrities and firefighters, a cook-off pitting firefighters against celebrity chefs, a trivia contest featuring professional athletes, and raffles with tons of prizes. I was invited about a month ago to act as the emcee, but I had no idea there would be so much going on. It sounds like a fun day, but I'll be doing a lot of speaking, almost as much as a radio DJ. I decide to dig up a bunch of fun trivia questions and also to watch cooking shows like the *Iron Chef* to get a better idea of what to say.

The rest of the afternoon seems to fly by. I enter the sound booth to lay down a few voice-overs, which need updating since the last time I recorded them. Sometimes information regarding a community, a home or an event changes, since we shoot three weeks or more ahead of airtime. I'm also responsible for doing quality control (QC) on my own show every week, which means I review the tapes to make sure things flow smoothly and there aren't any fatal errors in the piece. Most of the time I'm pleased with the way it turns out, although every now and then I suggest tweaking a special effect or adding a snippet of music in a particular place. Today I find five things that need to be fixed, which is disappointing. I also find myself a little agitated that the editors cut one of my stand-ups shots, translating into a loss of airtime. Five seconds may not sound like a lot, but we on-air personalities guard our time jealously. I e-mail my editors, asking for the necessary changes to be made, and then pack up for the day.

After work I meet my boyfriend for a nice dinner, where we talk over what's happened so far this week. He's very supportive, which is quite helpful to me. I know maintaining a personal relationship with someone in the entertainment field can be difficult, especially when one's career demands so much time. Creative-type people seem to be more sympathetic to the hard work that goes into being on television or acting in movies, but things can get awkward if you haven't worked for a while, or if your star starts to outshine theirs. I try to stay humble and remember that any of us are only a day away from being back on that dinner theatre stage.

LESSONS/PROBLEMS

Sometimes I over-commit myself to my work because of a strong motivation to attain as much public exposure as possible. This probably isn't the best week for me to host the charity event, what with all the turmoil at work, but it's something I really want to do. I have to keep in mind, though, that working six days a week can be a killer. I sure do miss my eight hours of sleep every night!

SHOTS FROM MY VALENTINE'S DAY SHOOT AT DELANEY VINEYARDS.

Day 5 JANUARY 19

PREDICTIONS
- Film my Valentine's Day show at the vineyard
- Select a new batch of clothes from my sponsor
- Meet with the charity people to plan tomorrow's event

DIARY

Things start off a bit crazy today. I arrive at work to discover that my DP has read the directions incorrectly and is already on his way to our shoot. I'm supposed to ride with him, so I ring his cell phone and ask him to return to the office. This puts us about 15 minutes behind schedule, so I also let my contact at the vineyard know we're running late. The weather is cold and rainy, which will make it a challenge to appear "glamorous" for our Valentine's Day show.

A friendly and helpful vineyard staff member greets us, helping me ignore my rumbling stomach—no time for breakfast this morning—and enjoy the tour they give us before we set up our equipment. This is the first vineyard I've ever visited. The huge barrels, the vast bottling facility, and row after row of grapevines fascinate me. Chris is my DP today, and we take some time to select several locations that will prove easiest to shoot and also the most interesting to our viewers. I interview the vineyard's CEO, asking him to briefly explain the winemaking process, and he does a very nice job of it. We're paired together on-camera—known in the business as a two-shot—while he pours me a glass of red wine and exclaims that his favorite thing about wine is sharing it. I close the show with a warm, "Happy Valentine's Day," and we clink glasses in a toast. The lighting inside the main building is low without being dim; it gives off the sort of romantic tone I had in mind when I thought up the idea of a Valentine's Day wine tour.

The DP and I shoot seven additional stand-ups in places throughout the vineyard. The last one we do will actually serve as the show's opening scene. Show "opens" are extremely important, since they're designed to immediately capture the viewer's interest and also explain where I am and what I'm doing. Chris decides we should shoot this open outdoors. Although our surroundings are beautiful, it's still raining and probably right around 32 degrees. Also, there's the wet ground to consider. Chris hands me an umbrella and we drag ourselves through the mud to find a good place to set up. Sadly I've worn my black suede heels today. By the time we reach our chosen location, my shoes are pretty much ruined. It's moments like these that make you realize no one watching the show will ever comprehend what hardships took place so we could capture this one shot. Before we're finished I'm cold, soaking wet, and my shoes are caked in orange-red mud. Just before we leave the owners give each of us a complimentary bottle of red wine, which is a nice touch.

After drying off over lunch, I retreat to the office to record my voice-overs for a show due this week. Our client list consists of builders, developers and managers who want their properties featured on our show, and they often pay for this privilege, a nice revenue generator for us. But clients often suggest last-minute changes to a segment produced several weeks earlier, which is why I'm recording some revamped material today.

Once I'm finished with that, it's time to head to my clothing sponsor and return this week's wardrobe. My personal shopper is busy with another customer, so I decide to select my own wardrobe for next week. I roam through the store in search of well-tailored clothes that will come across attractively on camera. My wardrobe rules involve selecting brightly colored outfits—nothing garish, of course, but certainly something that offers what I call "pop." I love the way royal blue looks on camera, and one workshop I attended a few years ago explained that people tend to respect you more if you wear that particular color. Royal blue rates as a "power" color, but non-threatening in a way that bright reds or yellows are not. I pick four outfits and check them out with the clerk before driving over to the charity event meeting.

We take two hours to discuss how various events will be staged tomorrow. Working as an emcee for a live event is much more challenging than anything I do on camera, since there are no "do-overs." The staff offers me a clear understanding of what's expected of me, and I secretly start to freak out because I haven't done enough prep work. I know I'll be interviewing several local and national celebrities and presiding over a cook-off that will pit celebrity chefs and a group of firefighters against each other. I spend the rest of my evening writing out note cards, researching trivia questions, and investigating the *Iron Chef* television show. I'm admittedly a bit nervous about tomorrow.

LESSONS/PROBLEMS

One factor I always forget is that I have no control over the climate! I should be better prepared for inclement weather and remember to pack an umbrella for rainy days, a winter hat for bad-hair moments, and tennis shoes if it's likely I'll have to trudge through mud to shoot a stand-up. Also, committing myself to emcee an event for which I did not have time to prepare was probably not such a good idea, but managing my time better would probably solve that issue. I simply can't afford to turn away such good opportunities for exposure.

Day 6 JANUARY 22

PREDICTIONS
- Catch up on office work
- Obtain footage for my demo reel from other productions I've worked on this year

DIARY

I head for the office, tired from this past weekend's events but proud of my accomplishments. Saturday went better than expected, despite the daylong rain and freezing temperatures. I was bundled up really well all day but the cold caused my hands and feet to go numb, making it a challenge to hold onto the microphone. The local media was on hand to cover the activities, which allowed me to give the newspaper a quick interview and plug my show. Sometimes you have to be a shameless promoter! I found that the note cards I'd written out really helped when it came to interviewing celebrity guests. My former boss did some filming, so I'm hopeful I can use some of the footage for my demo reel.

Morale around the office is still lower than normal following last week's layoffs. Our corporate office sent out e-mail to announce a dessert-based happy hour for tomorrow, which means that someone noticed we weren't our usual cheerful selves. I have no shoots scheduled for today and I've already picked up my clothes for this week's shows, so I'm free to sit at my desk and finish some overdue paperwork.

I've been asked to produce another "special feature," which is different from one of our standard show segments. Special features usually have a specific topic, like money or entertainment, or else they highlight a particular city that's interesting or exciting. I'm expected to produce them on a bi-monthly basis. These shows involve quite a bit of production work, but they aren't much fun since they require minimal on-camera time. In the past I've proposed several special features that I felt were more creative and humorous than those we generally run, but most of these ideas have been rejected. My frustration stems from the fact that our corporate office has been running this show the same way for nearly a decade, and their philosophy is, "If it ain't broke, don't fix it." Their reluctance to explore unfamiliar territory has caused the show to remain pretty much the same from one year to the next. I don't expect it to become controversial, but I'm a firm believer that variety is the spice of life. Some battles are worth fighting, but the recent lay-offs tell me this isn't the time to argue for more exciting content.

I brainstorm ideas for my special feature. Certain cities in the area come to mind, and it's been proven that our corporate heads like this angle, so that's what I'll propose. I also e-mail several production companies that have footage of work I've done. I'm adamant about putting together a new demo, and there's no time like the present to beef up my personal marketing campaign.

After a day at the office that includes surfing the Internet and eating handfuls of popcorn, I decide to visit mom after work. I try to spend as much time as I can with my family, especially since they're so supportive of what I do. I often use my mother as a sounding board for things that happen with the show. She records it every Sunday and watches closely to report on what we've done. It's nice to have someone who has seen your climb thus far and is able to give you positive feedback, plus the encouragement to aim even higher. We have a good talk about the direction things are going in, and she seems just as eager as I am about my press kit. Earlier today I learned that my press kit shipment is on its way. I can't wait to see it!

LESSONS/PROBLEMS

I'm having difficulty tracking down copies of all my previous work and actually getting production companies to hand over the DVDs. I shouldn't let so much time pass between when I do the shoot and when I call to obtain the footage. I also need to face the fact that a special feature is due soon, and I haven't yet done any production on it.

Day 7 JANUARY 23

PREDICTIONS
- Pick up footage for possible addition to my demo reel
- Visit my salon sponsor for a manicure and pedicure
- Make an appearance at the NHL All-Star skills competition

DIARY

I begin my day by narrowing down the city search for my special feature, and soon I find one that fits the bill. I send e-mails to the town's nature conservatory, its performing arts center and their main shopping district, figuring that the latter might have some interesting stories to weave into the narrative. Once that's finished, I respond to a lovely e-mail from the company for whom I emceed this past weekend. They were pleased with the way things went and suggested we work together in the future. I thank them for their kind words and let them know I'd enjoy doing another gig. Options like these make me feel more secure, in case I need some extra cash or wish to sharpen my skills in front of a live audience.

It's amazing how slow the day seems to go when I don't have any shoots. Sitting in an office just isn't my thing. I like the freedom of spending time outdoors rather than in a studio. I surf the Internet to generate ideas for my next show and for my upcoming special feature. Soon I hear back from both the conservatory and the performing arts center. Each venue is interested in being part of the show, so that's

A LITTLE RELAXATION TIME ON THE JOB HELPS FROM TIME TO TIME...EVEN IN MY ON-CAMERA WARDROBE.

very good news. I reply with our show's details and let them know I'll be in touch with a confirmed shoot date as soon as we coordinate our schedule.

Then it's time to call the final production company holding footage for my demo reel. The gentleman is happy to copy the material onto a DVD but "regrets to inform me" that the sound had failed during the taping. "You look great," he says, "but there isn't a lick of audio." How could that have happened? I think about it for a few seconds and realize that even without audio, if I look my best, I can always use the footage as part of a montage with voice-over narration. I decide to have them burn the DVD and I'll sift through the bad stuff to find fun, energetic moments that won't require sound to appear exciting. Glitches sometimes happen, no matter how well the event is staged or how successful the production company may be. You have to learn to roll with the punches, and today is one of those days.

I'm relieved to find my afternoon open, which allows me to visit my salon sponsor. This place cuts and colors my hair and gives me a manicure every four weeks in exchange for being listed in the show's credits, as well as my promoting them at various live events we do. Most shows do this. It's a win-win situation for the provider and the on-air talent, and this arrangement has saved me tons of money on grooming expenses. My sponsor has been featured on quite a few national television shows for their work with celebrities. I've waited far longer than necessary for a manicure, because of my busy schedule, so I include a pedicure with everything else. As odd as it sounds, if you're a woman and wear open-toed shoes, people pay attention to your feet. The salon is not crowded, so I get seated right away. They start work on my hands and feet simultaneously, which saves me a lot of time. I try to relax during the process and breathe out all the stress, anxiety and millions of stray thoughts that roll around in my head. I always try to use this time to think about nothing, which is perhaps the greatest luxury.

After my fingers and toes are dry, I rush home to change clothes for the second evening of NHL All-Star festivities. Tonight's featured events include a skills competition among young athletes. I enjoy returning to the American Airlines Center—where the Dallas Stars play hockey—to visit with old friends. When I go for popcorn during one intermission, two young men approach and ask if I'm the host of *that* home show. It gives me a chuckle, especially since teenagers are not part of our show's expected demographic. It's both flattering and weird when I'm recognized in public, but there is another side to that coin as well—being disliked for appearing on television. Since I come into people's homes every week, they feel as if they know me, which can lead to criticism. We do it with celebrities all the time. We don't know them personally, but we say whether or not we like them, commenting on what they wear, how they act, and whom they date. Whenever someone has something positive to say to me, I try to express my deep appreciation. After all, there can always be a darker alternative.

LESSONS/PROBLEMS

I need to figure out a way to incorporate the no-audio footage into my demo reel, while still making it relevant. I'll also have to compile a comprehensive list of media contacts to receive the demo, once it's ready to go.

MY DESK, WHERE I SCRIPT, WRITE SUP'S, AND MAKE REMINDER CALLS.

Day 8 JANUARY 24

PREDICTIONS

- *Film at a community that is rather far away, which means I may miss my lunch hour*
- *Pick up the remaining DVD footage I'll need to complete my demo reel*
- *Attend the final event of the NHL All-Star week*

DIARY

It's another cold, rainy day, and I'm somewhat tired from staying out late to watch last night's All-Star skills competition. My first shoot is set for 9:30 A.M. in a location about an hour away. It's marked as a new shoot, meaning we have no previous b-roll (background) footage of the location.

Our cameraperson will need to capture all that, plus film the interviews as I do them. When we arrive, there is some confusion with the man I'm supposed to interview. He claims we're here on the wrong date, but rechecking my schedule seems to refute that claim. I call my office and speak with my scheduling assistant, telling her the gentleman is not prepared and the other person I'm supposed to interview has taken the day off! She double-checks the schedule and also calls up their marketing coordinator. Several minutes later, she calls me back. Sure enough, I was correct all along.

When things like this happen, I'm not sure if there really was a mix-up or if the people I'm scheduled to interview are merely trying to avoid it. For every person who wants to be on television, there are at least two others adamantly against it. This guy seems nervous, and I'm almost ready to believe he made up the entire scheduling issue. As silly as it sounds, grown men and women can be deathly afraid of the camera, even to the point of throwing up, crying, having a fit, or just being plain difficult. I've interviewed people who sat with their arms crossed and provided grudging, one-word answers to my questions, and there have even been people who cursed at me because it was my job to interview them. It's amazing how that tiny camera lens can rattle someone's emotional cage, but it does.

It takes me a bit longer to get things going, mainly because today I'm using a brand-new recording device that came without an instruction manual. However, I eventually get my equipment going and then have my interviewee, whom I've secretly labeled Mr. Nervous, sign a waiver so we can begin. I start with basic questions about the community, such as a description of the location, what sort of people live here, any special amenities, and so on. I'm surprised by the ease with which he's able to respond. Considering how nervous he seemed at the outset, I fully expected this shoot to be a nightmare. As we continue, it's obvious that he's had some media experience. His responses are clear and concise, exactly what I want! He's probably better than most people I interview on a daily basis, which piques my curiosity. I ask if he's ever been on camera, and he looks at the floor and smiles shyly. He turns out to be the father of a young girl involved in a high-profile accident back in the 1980s. Although I was quite young at the time, I clearly recall the tale and the made-for-television movie that told her story. But he volunteers no additional information, so I let the issue drop.

Later I return to the office to script this morning's interview. Scripting is a piece of cake when you're working with a media-savvy subject. They keep their responses short—in other words, no run-on sentences—and know to smile pleasantly when they're on camera. Since this operation takes less time than usual, I make a call to see when I can pick up my demo reel footage. To my surprise, the producer had already sent the DVD to my building via courier, which I retrieve from the front desk. I finish up my office work; double-check with my managing producer regarding tomorrow's schedule, and head home to take a peek at the DVD. The film is upbeat and energetic, terrific footage for my demo reel, except for the fact that there's no audio. This was a national simulcast from New Year's Eve so I was wearing a fancy outfit, my interviews were comfortably informal, and the crowd of around 700 enjoyed three DJs who spun tunes all night long. This segment offered a much different look than the way I appear on my show, which is the whole point of including it here. I'm disappointed about the lack of sound, but I sift through the footage to find action shots that won't need audio to capture the essence of what I'm doing. I find three short segments, which I guess will have to suffice.

I get dressed for the final game of NHL All-Star Week and head out for the night. When life is as busy socially as it has been this week, I skip the gym and vow to be especially careful about what I eat. Oftentimes it's more important to be out and about, networking with others, than to get bogged down in the intricacies of a daily routine.

LESSONS/PROBLEMS

Today I learned that I'm developing the ability to bring out the best in people on camera, even if they're reluctant to be interviewed. I also realize I'll have to find a way to turn the disadvantage of my soundless New Year's footage into an advantage for my demo reel. That won't be easy.

I SIT QUIETLY TO WRITE AND PREPARE MY SUP.

Day 9 JANUARY 25

PREDICTIONS

- There are two shoots today, which means I'll be very busy
- My clothing sponsor needs my outfits back so they can do an inventory count

DIARY

I leave for work later than usual this morning, since the first shoot is set closer to home than my office. I let my DP know I'll meet him there and then make a quick stop at a donut shop. Though I try to watch what I eat, this busy week has driven me to seriously crave donuts. I eat while I'm driving and simply zone out by listening to the radio. I have a nagging question about the jury summons I received in yes-

MORRIS SETS UP THE EQUIPMENT TO SHOOT MY SUP, MAKING SURE THE LIGHTING IS CORRECT.

terday's mail. This is the second time I've been selected for jury duty. The first time I was excused on a hardship claim because I was self-employed and the trial was scheduled to last several weeks, but now I'm working for a corporation. By law employers are obliged to let you serve as a juror when called, but I'm not sure what to do in my situation. Since I'm the only host of the show and we shoot five days a week, I have no clue what missing that time from work would do to both my job security and getting the show produced. It's not like I have a replacement waiting in the wings, certainly not one that could be worked in without raising a red flag to my viewers. I decide to ask the court for a postponement in a letter that explains my unusual circumstances.

When I arrive at our location, my DP and our grip pull up

right behind me. I help them unload the van and th[e]
filling out all the waivers I'll need today. The marketin[g coor-]
dinator for this particular shoot has scheduled more pe[ople]
to be interviewed than we usually allow, but I agree to se[e]
them all. To conserve time we typically schedule no more
than four interviews per segment, unless it's a party-type
event. Nearly everyone here seems excited to be on television, although one man among them seems a little deadpan. I make a mental note to work on him. Unless you're a hard-hitting reporter interviewing victims of a tragedy, it's a good idea to have happy, smiling faces in your segments. People are drawn to others when they express joy, laughter and warmth on camera, but those being interviewed don't often convey those feelings if they're a bag of nerves. Once the deadpan gentleman sits down for his interview, my DP has to constantly remind him not to play with his microphone cord, to look at me and not directly into the camera when giving his answers, and to liven up. This man is much older than the rest of his coworkers and seems stuck in his ways, so our suggestions don't do much good. There are two reasons why this is bad. First, it makes the man look like a piece of talking Styrofoam. Second, it makes my segment look and sound boring, as if I'm unable to craft a fun piece.

I come to grips with the fact that things aren't going to get any better, no matter how much I smile and look directly into his eyes. We move on by shooting the rest of the interviews, which go smoothly. Once my subjects have cleared out from the gorgeous model home that is today's set, I begin to compose the show's open and close. The DP and the grip discuss our lighting—it's a sunny day—and set up our open in the spacious living room. I'm wearing a dress that stands out against the muted colors of the home. Even though I'm not doing any fancy intros or special effects, I'm sure my outfit will be a major motivator to make channel-flipping viewers stop and see what I'm doing.

Once our shoot is a wrap, we three head for a quick lunch before making it to our second shoot of the day. I'm glad this will be a short assignment, with only one person to interview. The community where we're shooting is one we've done before, which means we only need to capture the interview on tape. All the background material we'll need has already been done. The interviewee is not only well prepared, but smiles frequently and repeats back the question I've asked as part of his answer, which is incredibly helpful when scripting a segment. We're in and out in less than an hour, and I can hardly believe my good fortune.

Now it's time to drive across town to my clothing sponsor. This weekend is their annual inventory check, so my borrowed clothing needs to be returned so they can be counted. I drop off everything in my possession and return home early to rest and relax. I park myself in front of the television, eat some snacks, and check my e-mail. Then I force myself to head to the gym for a yoga session, which I know will help keep my head on straight, figuratively speaking at least.

LESSONS/PROBLEMS

I ought to realize it's not possible to make everyone look and sound good on camera. They have to want it for themselves. It's also hard for me to admit that some people just won't ever be good in front of a camera, no matter what the circumstances. One of my goals is to devise a few tricks so my segments will look and sound exciting, even though I may be stuck with a bad interview subject.

Day 10 JANUARY 26

PREDICTIONS

- Clean out my office in preparation for my move to corporate headquarters
- Check with a producer interested in having me host an as-yet undisclosed project

DIARY

Today's good news is that it's Friday, and the weekend is almost here. The bad news comes in the form of a phone call from my managing producer, who tells me our company will no longer retain a separate production office. This is startling, since we have a beautiful place where I have my own space, my own voice-over booth, and a lot more privacy than the corporate facility offers. But with the employee layoffs, it apparently doesn't make sense to keep an extra office when thirty spaces have opened up in their building. I try not to let this new revelation bother me, but it's obvious my fellow workers feel equally put-upon. I drive to the production office one last time to pack up my belongings. Everyone else is doing the same, clearing their spaces and tossing out memorabilia. Once we arrive at the corporate office, our managing producer shows us to our respective spaces. We're occupying small cubicles in an open office, lit by overhead fluorescent lights. It's a far cry from the comfort of our old place, but there's no point in saying much about it. Most days I'm on the road, so fortunately my actual office time is minimal. It does worry me that our layoffs haven't entirely ceased. At the very least I'm motivated to finish my demo reel.

After making the big move, I meet my mother for lunch at a nearby mall. We've spent very little time together lately, so it's nice to see her again and catch up on news and events. I explain what's been going on at work, and she has a much calmer approach about it. Since this is my first experience with corporate downsizing it's possible I'm overreacting, even though I feel stressed out. My lunch breaks are only an hour long, so all too soon it's time to head back to the office, record some voice-overs, and script a few last-minute items for this week's program. One thing many shows do is refresh segments instead of re-running them as-is or tossing them out entirely. I go back and add new sound bites to an old segment so it will appear new, even if the details or the subjects are essentially the same. You see this a lot with shows like ours, such as with entertainment news broadcasts that employ the same b-roll footage of a celebrity, even though the "breaking story" is new. Surprisingly, most of the time viewers don't even notice it.

The last thing on today's agenda is to check with a producer who wants to discuss a new project with me. A former director of an infomercial I'd hosted had shown it to a fellow producer, and it was this person who suggested we collaborate on something together. There's no answer to my phone call, so this will be a project for Monday morning. I'm already finding it difficult to discuss such matters in my new office without everyone hearing the whole conversation, and I'm especially careful not to discuss money in front of my co-workers. I've seen too many people discover someone else's salary or find out about their side gigs and a war of jealousy ensues. Consequently I try to avoid discussing my extracurricular jobs at the office.

LESSONS/PROBLEMS

It will definitely be a difficult adjustment being in a new office with a less attractive and considerably less private workspace. It's also going to be tougher to discuss side projects on the phone or via e-mail with so many co-workers close at hand. I can't allow the circumstances to deter me from seeking other work, however, since I need to stay sharp and vary my skills.

PREPARING TO SHOOT MY SUP MEANS RUNNING LINES SEVERAL TIMES FOR AUDIO AND LIGHTING CHECKS.

Day 11 JANUARY 29

PREDICTIONS

- Expect more office turmoil, considering our move
- Return outfits to my clothing sponsor
- Catch up with the new producer about our possible joint project

DIARY

It's another Monday, but this time I'm driving to the corporate office instead of the "old" production office. There is definitely some tension about, since we're crammed into a conference room until we get our new gigs sorted out. I struggle to get my office work done, which includes logging shoot information into the computer system, making changes to scripts, and e-mailing my contacts. I call each of my contacts for this week's shoots and leave messages that remind them of our start time. I enter this information into the computer along with my initials, just in case they try to pull the "I didn't know we had a shoot scheduled" card. I try to do this every week, but there are times when I'm too busy to call. After last week's slipup where the interviewee claimed that we weren't scheduled that day, I realize I need to return to this method.

I also begin to call my contacts for the special feature I've scheduled for Monday, reminding them of our plan. I leave each one a message and begin to map out my interview questions. These two locations include a busy downtown shopping/dining/recreation area and a performing arts center. I've never been to either place, so I check out their respective Web sites and make note of some questions to ask them. I jot down, "Tell me about your building's history. Tell me about the upcoming shows you'll be presenting at the theatre. What makes this downtown area unique? What are some of your most popular attractions, and why?"

Lunch is on my mind after a morning of slaving over the computer. I meet a close friend for a quick bite, filling him in on the downsizing and related chaos at work, since I know he can relate to it. While not part of the entertainment industry, his company went through something similar about a month ago. He agrees that it's a good idea for me to start sending out my press kit as soon as possible. After leaving him, I feel better about the fact that my press kit has arrived at my house and will be there tonight when I get home. There is so much potential for where it can take me.

I decide to drive myself to the afternoon shoot rather than ride along in the van. Today's crew was on another shoot this morning, so it makes sense for me to meet them there. We're also shooting a new location. If I didn't have my own car, I'd have to wait around for them to finish filming rather than leave as soon as my portion is complete. On my drive over I call up the producer regarding the project he has in mind. We finally connect, and he tells me immediately that he definitely wants me to host it. He'd seen my last similar venture and believes I'll be a good fit for this new one. After explaining my hectic weekday schedule, we agree that evenings and weekends are our best options for this side gig. When I ask for details he suggests we meet over lunch to discuss it, but briefly describes the work as an "awareness" video for a wrongful-death investigation that has gained national media attention. This definitely piques my interest, so I let him know that lunch on Thursday will work, where we can discuss compensation and other pertinent details.

This afternoon I have four interviews to conduct. Unfortunately, my personal video equipment has not been properly charged. This isn't a major problem, but it means I won't be scripting today's shoot on location. Fortunately, each of the interviewees is prepped and ready to go, with their responses well thought out. I interview a realtor, a homeowner, a city official, and a home community sales representative. Each one is uniquely interesting, smart and funny, which makes me confident this segment will turn out well. We quickly re-set the camera and lights for my stand-up. I pull out what I'd written earlier and quickly memorize it. Then we shoot, and I nail it on the first take!

The benefit of having brought my own car kicks in, as I'm able to leave while the DP and grip finish their part of the shoot. When I arrive at my clothing sponsor's location, the manager has several things already in mind for me this week. Instead of trying everything on, as I usually do, I let her wrap it all to go. Before departing I tell her our company's CEO forwarded an e-mail message to me from a viewer who had raved about my on-screen attire. Naturally she is pleased to hear about it. The styles at this shop are very cutting-edge, the sort of items I wouldn't normally select on my own. But that's more because they're beyond my budget rather than unflattering or too modern.

I make it home in time for yoga, which I follow with a light snack dinner and a long look at my new press kit. It turned out well, especially the container that will hold my demo reel. The print quality is high and the DVD case gives off a really professional look. My photo on the front could be a little higher resolution, but I guess that's my fault for supplying a low-rez photo in the first place. It's no major deal, but I'm sorry I hadn't thought this issue through. Overall they're beautifully done and will definitely serve the purpose for which they're intended. Tomorrow I'll stop at my editor's place to begin selecting specific clips to be copied onto the DVD.

LESSONS/PROBLEMS

I learned an important lesson about paying better attention to the resolution of my photos when using them for high-grade print jobs. I also learned that charging my personal equipment ahead of time means saving time during the scripting process. The only lingering problem is not having a desk area at work.

Day 12 JANUARY 30

PREDICTIONS
- Participate in a shoot in the morning
- Work on my host allowance invoice for next month
- Begin selecting my demo reel shots

DIARY

I begin the workday by learning that this morning's shoot is not only some distance away from the office, but also a new location. Unless I care to hang around all day while my DP does his background filming, it's best if I drive my own car. Many TV hosts receive an allowance to cover the cost of makeup, wardrobe and other miscellaneous items, generally calculated on a daily (per diem) or monthly basis. For anyone who doesn't, it's best to analyze your salary to see if it's high enough to justify what you spend on fitness, grooming, and all those other elements that help make you the best at your profession. Costs add up quickly, such as the money I'm spending these days by driving my own car to all these far-flung locations. I fill out my expense form and forward it to my managing producer for approval. I've come to depend heavily on this amount each month to help pay my bills, even though having a fashion sponsor and a salon sponsor have eliminated the standard clothing, hair and nail expenses.

I follow our show van to the shoot location, which seems like it's in the middle of nowhere. I'm appalled by how long

IT'S ALWAYS USEFUL TO HAVE A GRIP ON A SHOOT.

it takes to reach it, mentally calculating how much gas I've already used this month. When we arrive there, the four interviewees are poised and ready for me to put them through their paces. I hook up my recording equipment with the hope that it's charged and ready to go. I check that it's displaying the proper time code so I can log each sound bite properly, and also inform my editors how long each sound bite lasts. Once I get this configured, I have the interviewees fill out their waivers so we can begin. It's yet another lucky day for me, interviewing articulate people who respond in pleasantly concise sentences to give me excellent sound bites. My cell phone rings while I'm conducting my interviews. When I check the message at an appropriate stopping point, it's my managing producer telling me a client of ours is unhappy with one of the segments we've produced. Even

THE SLATE HAS TO BE DONE PRIOR TO INTERVIEWS SO INTERVIEWEE INFORMATION IS MARKED ON TAPE.

though the material made it all the way through our QC process, with the editors and me finding it OK, apparently the client was less than pleased with the way it turned out. When I go back to the office this afternoon I'll re-shoot the show's introduction to make everyone happy.

On my drive back to the office I receive a call from a good friend I haven't seen in a while. She's someone I've known since my earliest days in the business and is still trying to break into it. For a while she was content to work for various marketing and promotional firms, but now she wants to be a news anchor. I'm convinced that taking a slightly longer lunch won't hurt, so we meet up at a restaurant back in the city. I wish I had better connections or advice to offer, but it's a difficult position to be in when someone your age and type wants help getting into the business. It's almost like robbing yourself if you give them free advice or access to your contacts, no matter how much you like them. I give her as many solid ideas as I can, such as finding an agent, having new headshots taken, and gaining more experience in her chosen field. She's a wonderful friend and I wish I could do more, but I'm in the process of exploring my own options.

Back at the office I script this morning's interviews and fax over the script to the editors. They take my written material and piece the segment together, one sound bite at a time. They match up b-roll footage to be shown on screen with whatever the interviewee is talking about. I keep a printed copy of the script and the waiver form in case I need to refresh a segment by removing old sound bites and replacing them with new ones, or in case someone tries to say we didn't have their permission to put them on the air.

Things are still quite tense around the office, so I'm definitely ready to leave work when the time comes. With fewer employees around, those who've remained are under greater scrutiny. I head over to my friend's house to begin downloading my demo footage to his computer. We work our way through several DVDs before realizing that some of the footage was not recorded in high-quality (HQ) format. That's not good. The picture looks fine on a computer screen, but on a TV it becomes highly pixilated, which means you can see the individual pixels that comprise each image. We agree that it's better to have our duplicating done in HQ—even though that means a delay—rather than rush through the process with lesser-quality footage.

Once I return home, I take another look at my press kit. The DVD case has a lovely, bright printed case slip with my photo and name on it, and the back contains a synopsis of my recent work. I have my photo printed on the DVD with a quote from a magazine that did an article on my former job with the Dallas Stars. This isn't a traditional press kit, but I believe it will encourage people to take out the DVD and watch my demo. I am hoping that the demo will lead to interest in my job, which will lead to press.

LESSONS/PROBLEMS

As before, I've learned the hard way that obtaining high-quality footage is not as easy as it seems. My two choices are obvious. Either I'll need to re-dub these existing tapes in HQ on DVD, or else wait for newer footage to air and use that instead.

AT PARTIES, THE SHOT IS SET UP FOR INTERVIEWS TO BE DONE OVER-THE-SHOULDER STYLE.

Day 13 JANUARY 31

PREDICTIONS

- Attend scheduled party shoot, though there is a threat of snow
- Participate in a side shoot after work, hosting a show that airs twice a year

DIARY

The atmosphere at the office continues to be tense, with everyone acting like they're walking on eggshells. Clients are especially nit-picky this time of year as well, since winter months mean fewer events on which to focus, and consequently more time to think about the way I'm putting my show together. My new office space has turned out to be a bit more private than my last space, which is pleasantly surprising. Even so, I'm keeping my personal calls to a minimum and making sure to complete my projects before they're due, if at all possible. These layoffs have significantly increased everyone's workload, which leaves less room for error when scripting, editing and producing my show.

This morning I'm shooting a party event, so I bring in my curling iron to touch up my hair before heading over there. While in the office I attempt to reach one of the subjects of next Monday's special feature shoot, a man who's proven to be almost impossible to catch on the phone. I leave him yet another voicemail message and back that up with e-mail. I want to make sure he knows exactly what time we're showing up and what we expect him to do.

At 10:30 it's time to head for the party with our company relations person. Today's DP and grip were out there earlier this morning to shoot the b-roll and get things set up. On our way over it starts to sleet, slowing the highway to a crawl. I'm afraid the inclement weather will lower the event's attendance, which will negatively impact the visual part of my segment. Upon arrival we're pleased to find the event's coordinators incredibly welcoming and helpful. Despite the low turnout, they provide a delicious lunch for us as well as a couple of terrific interviews. It's one of those days where I can't seem to compose anything interesting for my stand-up. Finally I scribble down some pretty basic thoughts and call Chris, my DP, over to discuss our staging. We intentionally design my stand-up to look different than my interviews

MORRIS SETS UP THREE DIFFERENT LIGHTS TO CREATE A WARM GLOW FOR THE SHOOT.

so the segment will have some visual variety. We settle on a wide shot that shows off the party space, and my walking toward the camera during my narration. At my final mark the camera will zoom in and finish with a tight shot that blocks out the background. I affix my microphone and stand still so Chris can check the lighting to make sure I'm neither too pale nor too tan on camera. Once I recite my lines to check sound levels and he rehearses his camera movements, we shoot the real thing. I nail it in one take, but when we replay it on the monitor, the script I used can be seen in the shot. That's a shame. I go back to my mark and we shoot it over. This time I can't seem to get it right, and it requires three separate takes to make a clean stand-up. Finally, we're wrapped for the day.

Back at the office I take the rest of the day to script the interviews from the party and enter the pertinent data from the shoot into the computer system. When the clock hits 5:30, I jump in my car and head to my shoot for the other production company. The script for this production is on a TelePrompTer, which is one of the reasons I took the gig. Because my daily work no longer includes TelePrompTer, I try to get as much practice with one as I can. It's important for hosts to write and memorize their stand-ups, but it's equally important to read from a TelePrompTer naturally and with ease. Commercials and news programs are often shot using a TelePrompTer, which is another reason hosts should be comfortable with the procedure. When I arrive, the set is already pulled together and the lighting is set. I'll be standing up while hosting and I head to my mark that's already placed on the floor. This is my third time doing this show, so the production team and I know each other quite well. We toy with the speed of the TelePrompTer until it feels right, and I use that opportunity to read through the script several times. The cameraman lets me know when we're rolling, and off we go. I stumble over one word and mispronounce another, but those problems are easily fixed and the rest of the shoot goes off without a hitch. A mere twenty minutes after getting there I'm already out the door, another 30-minute show set for broadcast.

LESSONS/PROBLEMS

Weather definitely affects the process of a shoot, especially it's an outdoor event. But even indoor ones can suffer if traffic is so bad it prevents people from getting there. A well-attended event always looks better on camera.

Day 14 FEBRUARY 1

PREDICTIONS
- Have an early-morning shoot, all indoors
- Go to lunch with my sister to discuss my career
- Spend a rare night at home

DIARY

The morning brings with it dampness and fog, which is not the most appealing weather in which to film. Today Morris is my DP, and together we load up the van and head to our first shoot of the day, driving for an hour in the rain to our location. I'm glad we're not filming outdoors, which would make for an unpleasant setting. When we reach our destination, our interviewees seem confused as to why we're there. I'd interviewed these two people earlier, but my boss asked me to return for more usable sound bites. They don't look all that happy about it, even though I'd called and sent e-mails as reminders. I check with my office to make sure we have the right details and, sure enough, we're indeed scheduled to be here today. One of my subjects tells me he's in a hurry, so we get him seated and interviewed straight away. I blow through the Q-and-A, and the next gentleman receives the same speedy treatment. For my stand-up we choose a wide shot that shows off my complete outfit, which is important since it retails for close to a grand. After three takes we have two stand-ups in the can and we're ready to return to the office.

SPENDING TIME WITH MY SISTER IS A GOOD WAY TO DE-STRESS.

Now it's time for lunch with my sister. She works for a healthcare facility located only a few miles from my office. She's contemplating a career change, so most of our conversation centers on what I do for a living. I explain that being a television host is both rewarding and exciting, but the business has the same negatives as any other field. Work can be monotonous at times, depending on the specifics of the job, and we're subject to the same corporate whims regarding profit and loss. In general, I tell her, it's highly rewarding to see your show on the air, especially when viewers offer positive feedback.

After lunch it's time to go back to the office and print out maps, interview questions and contact details for my special-feature shoot scheduled for Monday. One of my contacts has confirmed via e-mail that he'll provide us access to several shops and restaurants in the shopping center, which will liven up the piece and provide some exciting b-roll. Then I script our morning's interviews and deliver the printed script and waiver forms in person to my editors. Editors are highly creative and patient people. After seeing what they do on this show I have a greater appreciation for how hard they work. It's always a good idea to be as friendly and helpful to the editors as you can, since they have the ability to affect how you look on camera.

Ever since the layoffs, everyone here is carrying three times the standard workload. With a few minutes left at the end of the day, I sit at my computer and come up with five things I like and admire about each person who's part of the production team. I print up individual copies and hand them out. It takes an amazing team to put on a quality television show. When people know they're appreciated, it makes coming to work and doing the job that much better. It's fun to see everyone smile for a change, and I feel better about concentrating on the positive aspects of the job rather than the negative ones.

Since the weather has taken an even nastier turn, I decide to make tonight a quiet one at home. It's funny how I'm directly involved in creating television, but I hardly take time to watch it these days. I miss seeing how other hosts conduct themselves in front of the camera. I learn quite a bit from watching them and find myself motivated after spending an hour or two understanding what makes other shows successful.

LESSONS/PROBLEMS

Today I came to understand the value of encouragement, even if it only boosts morale temporarily. I also learned that taking time for the people who help make you look your best is time well spent. Everyone involved with the production process is important. If they're not doing their jobs properly, I can't do mine.

THESE SHOTS REFLECT THE HOUR-GLASS ANALOGY THAT WE USED TO MAKE THE SHOW OPEN MORE FUN AND VISUALLY ENTERTAINING.

Day 15 FEBRUARY 2

PREDICTIONS

- Conduct a morning shoot, including three interviews
- Make last-minute preparations for shooting next week's special feature

DIARY

I'm hoping that today will be something of a catch-up day, once my morning shoot is finished. Spending so much time away from the office makes me feel like I'm falling behind in my work there. Snow covers the ground, so exterior shots will be tough to capture. Morris is today's DP, and we decide to film almost entirely inside the beautiful model home where we're conducting our interviews. While he sets up the lights and the camera, I start composing the stand-up scripts for the open and close. Trying to think up a clever angle, I spot an hourglass resting above the stone fireplace. It's a strong visual representation of time—the loss of it, the urgency of it, and so on. I decide to incorporate that theme into my stand-ups.

After Morris gets everything ready, I call in the three people I'm interviewing so they can sign their waivers. Once they're seated, slated and wearing microphones, I begin with my questions. Each of them does a fantastic job, offering me solid answers and friendly smiles. We wrap up quickly and re-set the scene for the show open. Several times I recite my stand-up out loud while Morris finds a new angle from which to film. We practice a few times on camera before recording it. When you're moving the camera a lot, it's best to do a few run-throughs, even if you don't feel like you need it. Sometimes I do it so the DP will be comfortable with what I plan to say. Morris focuses on the hourglass in synch with my phrase, "You'd better hurry; they're going fast and you're running out of time," a reference to the homes being sold in this subdivision. He closes with an extreme close-up of the grains of sand draining into the bottom half of the hourglass. We shoot the open a total of seven times, mainly to capture different camera angles and make sure that Morris hadn't cast any shadows onto my face. We set up one last time for the show close, trying to find an appropriate spot to showcase the hourglass as the remaining grains of sand fall through it. It's sort of corny, but the show close *is* about running out of time. Once again, this takes a little longer than expected because of the natural light coming through the windows and how it plays with the scene.

After a quick lunch I retire to my car for a much-needed 30-minute nap before returning to the office. Sometimes it helps to recharge like this before facing the rest of the day. After making the drive back to the office and scripting the segments we've just shot, I decide to e-mail my two contacts for Monday's special feature. With things already hectic, I don't need any unnecessary slip-ups. I'm delighted when both of them respond almost immediately. Now I can rest a little easier this weekend, knowing that we're all set for Monday morning.

At the end of the workday I sit at my desk and look through footage to see if there's anything else from previous shows that would make a great addition to my demo reel. While I'm making notes I receive e-mail from a production company for whom I shot a commercial a few weeks ago. They've included a link where I can view it online. Everything looks and sounds good, which pleases me. I have two separate demo reels on my professional Web site, one with commercials and the other with clips of my various hosting duties. This one will definitely go onto my Web site. Also, I try to update my photos on that site every six months. Most anchors, hosts and broadcasters have a tendency to alter their appearance slightly every six to 12 months. It's always a good idea to keep up with the competition.

LESSONS/PROBLEMS

There are times when I need to let the DP shoot more takes than I'd like, even if I've already found a stand-up that works for me. It's important to be sensitive to the way other people want to do things, even if doing so takes a little extra time. Even though it's my face in front of the camera, that doesn't mean we do everything my way.

THESE ARE STILLS FROM THE SPECIAL FEATURE SEGMENT I SHOT WITH CHRIS AND MORRIS IN THE OLD SODA SHOP. WE EVEN USED A VINTAGE JUKE BOX IN THE SHOW!

Day 16 FEBRUARY 5

PREDICTIONS

- Shoot my special feature
- Visit my clothing sponsor to return clothes and pick up more

DIARY

The morning begins with me feeling a bit under the weather. On the way to work I pick up some cold medicine to help me through today's shoot, since it's the day we're doing my special feature. I quickly fix my hair at the office and jump in the van alongside Morris, today's DP, and Chris, who will be our grip. Our first location is a prominent performing arts center. I meet with my contact to discuss where inside the center we can shoot our interview. He directs me to the main theatre area, which is gorgeous. This particular center is a historic landmark, with much of the interior restored to its original grandeur.

I give my subject a quick briefing as to the questions I'll be asking. He has a very dry sense of humor, and it takes me a few minutes to understand whether he's being sarcastic or serious. A host has to learn how to read people. Some prove more difficult than others, especially when they're animated, friendly and comfortable off camera but clam up when that little red light goes on. This gentleman seems very relaxed, which causes me to believe he's only joking around. The interview goes quite well, even though both Morris and Chris interrupt to suggest topics we should be covering. I know they're just trying to be helpful, and normally I don't mind this sort of thing. But today it irritates me. Over the past few weeks I've worked really hard to research this feature, making all the necessary arrangements and developing a complex picture of how I'll tell this story. Today I'm not in the mood for suggestions unless I ask for them, no matter how relevant they might be. After all, I don't tell these guys how to perform their jobs!

Once we wrap the interview we head for the balcony to shoot my stand-up, capturing the stage area below as background. It's a beautiful and interesting shot, and I'm pleased with the way it turns out. Walking through this historic area in search of a place to eat lunch, we spot a shop with an old-fashioned soda fountain and an authentic Wurlitzer jukebox. After a quick discussion, Morris decides this will be the perfect location for the segment that will open my special feature. We order our food and size up the rest of the place; it literally feels like we've traveled back in time. Chris feeds some quarters into the jukebox and we chow down to the sounds of Johnny Cash.

All too soon it's time for us to start our afternoon session. My contact is the PR person in charge of the entire downtown area, and she's lined up shops and restaurants willing to be part of this segment. Morris sets us up in an open space with U.S. flags in the background. He lights them beautifully, which gives the piece a very "all-American" feel. We have difficulty with this woman's interview. She interrupts herself, giggles nervously, and flubs a few words. I try to wrap things up quickly while still capturing enough sound bites to make it worthwhile.

She seems as relieved as I am when we finish, and she offers to accompany us while we visit the other locations she's provided. Our first stop is a restaurant with a beautiful patio. It's the perfect time of day to shoot, since the teeming lunch crowd is still in place. We greet the owner and obtain permission from the patrons to capture their faces in various crowd shots, and then Morris begins shooting his b-roll. Our next stop is a candy store that makes homemade candied popcorn and sells soft drinks in glass bottles, the old-fashioned way. The owners allow us to shoot unlimited b-roll of their store, and they even give us two giant bags of their delicious popcorn! I purchase a bottle of Coke and start to snack on our free treats. Today isn't turning out so badly after all.

Our final stop is a store selling antiques, which displays some incredible chandeliers and lovely refinished furniture. I have nothing special to do here, so while Morris shoots b-roll I'm already scribbling notes for the segments that will open and close this feature. Morris finishes quickly and then suggests we shoot outdoors. I really despise that, unless it's absolutely necessary. Outside it's nearly impossible to control surrounding elements such as traffic noise, people wandering into your shot, and the wind and weather. But rather than argue with him I agree to do as he suggests. Some days it simply isn't worth fighting with your DP. We shoot it as he wants, even though it takes quite a few tries for me to get it right. Once it's in the bag, we walk back to film our open at the soda counter. As a prop Morris orders a giant chocolate shake with whipped cream and a cherry on top. I sit at the counter and we run through several takes. There's one I'm especially pleased with, so mentally I'm finished for the day. However, Morris insists that we continue to work on it. I feel the frustration welling up inside me. I'm reminded that different people have differing ideas of perfection, and there are times when I have to accommodate them. I do one last take with all the power I can muster, and Morris gives me the high sign that tells me we're done at last!

LESSONS/PROBLEMS

When my co-workers give me advice on how to do my job better, it's a problem. If I discuss this with my managing producer he might mention it to the very people about whom

I'm complaining, which would make for an uncomfortable work situation. But if I say nothing, then they'll assume it's OK to interrupt my work and make their own suggestions. While I don't want to play the diva role, I still want the freedom to perform my job without excess commentary. Is that too much to ask?

Day 17 FEBRUARY 6

PREDICTIONS
- Attend a party shoot in the morning
- Script my interviews
- After work, coach a children's drama rehearsal

DIARY

For some reason this month has been chock full of party shoots, and there's another one this morning. I'm still under the weather, so I keep taking over-the-counter medication and drinking plenty of herbal tea. While checking my e-mail I see there's something urgent to fix before I leave for my shoot. One of my editors has identified a script that needs twenty additional seconds of sound. I have no idea how we came up short. But rather than ask a million questions, I find the DVD with the necessary sound bites on it and fast-forward until I find a good one. I type the time code for the sound bite onto a scripting spreadsheet and make sure to include the person's name, title, and the date on which it was shot. I also provide the tape number so the editors can locate it in their files without difficulty. Then I print out the spreadsheet for my own file and e-mail it to the three editors. Unfortunately, this is not the only urgent matter of the morning. Since our editors are behind in their work due to staff reductions, it's up to me to QC (quality-control) my show this morning. As soon as I provide them with my QC notes, they'll be shipping the show out for programming. Normally we have a few days to make the necessary changes, but no longer. With only thirty minutes left before my call time for this morning's shoot, I sit down to watch my Valentine's Day show. Everything looks fine until the end. Instead of the fully scripted close, my editors have a version cut nearly in half to 12 seconds. I'm seen starting the segment in mid-sentence, which looks ridiculous. I e-mail the editing team and my managing producer to protest this issue. I'm afraid we're so close to deadline that it won't get fixed, which of course defeats the entire QC concept. I get an immediate response from one of the editors, notifying me that *all* show closes will now run a maximum of 12 seconds, not 22 seconds as before. This is not good news. How can you sum up a program in 12 seconds? That's barely enough time to announce my name and say goodbye. I decide to meet with my managing producer soon to discuss it.

It's 58 miles to the site of the party—an unbelievable distance—and I arrive to see that Morris and our grip-for-hire have already shot community footage and the all-important b-roll. I try my best to mask my severe cold, but all four of my interviews turn out rather lackluster. Since there is a pretty serious party going on in the background, we decide to use that for my stand-up and have me walk toward the camera while I'm describing it. There's no time to write anything down, so I tell Morris he'll have to shoot me on the fly. There's quite a crowd behind me, listening and watching as we prepare to shoot, so I vow to do this in one take and not look like a total idiot in front of everyone. Morris rests the camera on his shoulder and I use the handheld microphone. I'm not convinced my dialogue turns out all that great, but at least we manage the scene in a single take.

Normally I like to schmooze with people at these events, but today all I want to do is have a bite to eat by myself and check my phone messages. I call back my agent, who tells me she's received a call about a commercial someone wants me to shoot. We both recognize the biggest obstacle will be finding a time that will work for the production company and me, since my show schedule is so booked up. I've also received calls from two other producers. One works for a major network with whom I've done several projects and he wants to know my availability for a shoot on Monday. That's unfortunate, since I have a scheduling conflict that afternoon with my own show. I tell him I'll be available if they advance the call time by just one hour. He's not too optimistic about it, but says he'll check with the crew and let me know. The other producer wants me to audition next Wednesday for a recurring gig that involves hosting a series of Web-based videos. This could translate into at least six months of side work, so I agree to come in and read for him. Everything will be on TelePrompTer, which works out nicely since I just did an assignment like that last week.

The last call I make is to the man who's producing that project on wrongful death. We finally nail down a time to meet in order to discuss compensation and the details of the project. With all these irons in the fire, something is bound to work out.

Back at the office I have barely enough time to grab my laptop before heading out the door. Every couple of months I volunteer to work with a drama team at my church. I enjoy using my skills for a good cause and giving something back to the community. Our theme this week is about helping others without expecting anything in return, a good reminder to me during a time when I'm so focused on getting ahead in my career.

LESSONS/PROBLEMS

I'm frustrated that my show closes will be chopped to nearly nothing, so I need to sit down with my managing producer tomorrow and figure out if there's an alternative. Also, the QC process needs to take place a whole lot sooner than the day the show is due. Without that cushion, there won't be any time to fix what's wrong or come up with changes that can make the show better.

Day 18 FEBRUARY 7

PREDICTIONS
- Spend the entire day at the office to catch up on my work
- Perform at church after work

DIARY

I'm excited I'll be in the office all day, which is a welcome change from my hectic shooting schedule. All the DPs are out on shoots that don't require hosting, so the office is quiet and things seem fairly calm. After responding to e-mail, it's time to log in all the necessary information into the computer system for this week's show. I've also been asked to refresh two segments, which means pulling out old scripts and adding new sound bites. I head to the editing deck and flip through the cataloged DVDs of recently recorded material. Once I find the exact interviews I need, I sift through them to find sound bites I can use to refresh the script. These sound bites are logged onto a spreadsheet and sent off to the editors. This is easier than scripting an entirely new segment, though it still makes the show seem fresh.

In between my job-related duties I glance through the script I'm using in tonight's children's drama. I have a surprisingly large number of lines, so it's vital I commit them to memory before I arrive. It's funny how memorization becomes second nature when cue cards or a TelePrompTer aren't handy. I can read through dialogue once or twice and remember up to 90 percent of it, a great skill to have when it comes to auditioning for projects, whether they're host-related or not. Being forced to memorize in this way has made me sharper and more competitive in my field. The topic of tonight's program is service, and how helping others without expecting anything in return is what we should all strive to do. I laugh at how this concept is so alien to my profession, where few people are willing to help you get ahead without expecting something in return. The entertainment business is full of shady characters, and you have to be careful whom you trust.

After a quick lunch with a friend and a pharmacy visit to pick up a prescription for my ever-worsening cold, I return to the office with the realization that I've completed most of my work for the day. I guess now would be a good time to organize my desk, something every host needs to do. Many times I've found that having my business card collection organized and close at hand leads me to just the right contact or gives me a terrific idea for the show. I also keep notebooks full of copies of each show's script in order to see what's already been done and what might be fun to bring back. Then there's my notebook that accompanies me on shoots, containing extra copies of signed waiver forms, a calendar of each day's shoots—when, where, what specific things we need to cover in the interviews—plus show ideas. Some days I find myself in a bind as far as creative writing goes, so I can always refer to this notebook for show opens, closes or introductions.

To kill time today I decide to devise a few new ideas for upcoming shows. One thing I've always wanted to try is some sort of circus theme. Chris, one of my DPs, worked part time as a clown back in his college days. He knows how to juggle interesting things, like fire sticks and machetes, and he can also ride a unicycle. In keeping with the circus-type theme, Cirque de Soleil often comes to town with new productions, and I figure it might be cool to shoot my show open and close at the event they always hold for members of the media. Holiday themes are a no-brainer, with nearly every show doing them, even broadcast news. Months like March and August are tough, so that's when the circus theme might come in handy.

Once I type out a list of potential show ideas, I file them away in my notebook for future reference. It's almost time for me to head out, so I review my lines one last time and make sure all the parts of my costume are packed in my bag. The evening turns into a fun and lively one with no missed lines and no technical errors, a flawless performance.

LESSONS/PROBLEMS

Having some down time recharges my physical batteries and allows me to become more organized for my next round of shoots. I also remind myself that what I do is supposed to be fun for lots of reasons, and not all monetary ones. Giving back in ways that have nothing to do with money, press or glory is the best thing about having talent.

Day 19 **FEBRUARY 8**

PREDICTIONS
- Do a morning shoot for sure, plus possibly one this afternoon for stand-ups
- Work with a new DP, which means things might take longer

DIARY

It's a rainy day, so I know we won't be doing any outdoor work today. The wind could cause audio problems as well, so we'll definitely need to come up with a solid place to shoot indoors. I arrive to discover that our substitute DP has not yet arrived. Ten minutes later, and still no DP has shown up. I don't have his contact information, so I call my office, hoping he's not lost. Just as I get through, he pulls up alongside me in our show's van. But our day's problems are only beginning. To my great surprise, the door of our shoot location is locked. I check my watch, and it's way past our scheduled shoot time. Then I look up the contact phone number in my trusty notebook, but the phone rings and rings so I leave a message. Finally I call my office again to let them know we're stranded in the rain, with no contact in sight. Our scheduling assistant calls their corporate headquarters, and we finally make some headway. The shoot appears to have been cancelled, but no one bothered to let me know!

By now I'm wet, cold and frustrated. Since we're paying this substitute DP a day rate, it's up to me to figure out a way to use him and make it worth the company's money. I'm not

THE MONITOR AND CAMERA ALLOW ME TO WATCH MY SUB IMMEDIATELY AFTER WE SHOOT IT TO SEE IF IT WORKED.

too happy about that, but now I have to find a place for us to shoot indoors while working with someone who's not my usual cameraperson. Don't get me wrong; this DP seems like a very nice guy, and I'm not going into this situation expecting the worst from him, but it's more difficult for me to work with someone who doesn't know my style. After a few calls I find a location that's willing to let us shoot today, so off we go. Upon our arrival I help him unload the light kit, the audio kit and the camera equipment before getting settled inside. While the sub starts to set things up, I get busy writing the stand-ups I'll need to complete this week's show. Meanwhile this DP is going awfully slowly. He tells me it'll only be ten more minutes, but that turns into twenty, and then thirty. I'm used to working with Morris or Chris, who do things far more quickly. Once he finally reaches a point where we're ready to start shooting, I sit down so he can adjust the shot and the lighting. When we start rolling, I nail the stand-up on the very first take. He asks me to do it again. I oblige, knowing that he probably wants to practice his camera movements a few more times. By the seventh time I recite this stand-up, I'm beginning to feel agitated. It's not really his fault. I know he wants to do the best job possible so my bosses will hire him again when there's another opening. But I'm used to a much faster pace, and slowing down to match someone else's speed is not my strong suit. I guess now I know how my regular DPs must have felt when I started working with them. I'm sure they had to operate more slowly with me, which more than likely cramped their style.

It takes us two hours to complete two stand-ups, work I usually finish in 30 minutes. By now I'm exhausted and can't wait to go to lunch. I usually help Morris and Chris pack up the lights and audio so we all can leave more quickly, but today I decide to be less helpful and simply book out of there.

After an abbreviated lunch break and the hour-long drive back to the office, there's more bad news. Our substitute DP had returned earlier with tape from our morning's stand-ups, and my managing producer doesn't like any of it. He considers all the shots under-lit and essentially unusable. Those two hours of torture went for nothing and the company wasted money on his day rate after all. This means that tomorrow, a day already booked solid with shoots and appointments—including one at my sponsoring salon—we'll need to re-shoot these stand ups!

I spend the rest of the afternoon trying to figure out how to balance all of tomorrow's commitments. In the morning we're doing show promos in the studio. In the afternoon we've scheduled a new location with four interviews. Add to that these re-shoots and my monthly hair appointment, and it's an overfull day. Unfortunately the afternoon shoot won't be scripted until Monday, and my voice-overs won't be recorded in time to meet the deadline. I speak to my managing producer, and he agrees to make things easier by moving tomorrow's studio session to Monday morning, which will give me time to get organized and collect my thoughts. Thankfully my regular DP, Morris, will be shooting tomorrow.

LESSONS/PROBLEMS

I know I need to remain flexible and accommodate other people's working styles, but I also want them to be understanding of mine. I had a problem with how much slower this day's shoot went considering what I'm used to, but I need to be patient with people who are new to various aspects of the business. I was once in their shoes, and I know I will be again.

Day 20 ~~FEBRUARY 9~~

PREDICTIONS
- Re-shoot yesterday's stand-ups
- Shoot a show's open and close
- Keep a hair salon appointment

DIARY

I start out knowing it will be an incredibly busy day. Our first location is at a master-planned community that features swimming pools, golf courses and lakes. These amenities will make it much more fun than a standard indoor shoot. When I arrive at the location, Morris and the grip are already finished with the lighting and about to hook up the audio. I feel good today. I love my outfit and I know I'll be able to leave a bit early for some much-needed pampering at the salon. I sit down to write my show open and close as quickly as I can, and then Morris comes up with a great idea. We'll set up with an empty background, but while I'm talking items will magically appear behind me. This is an easy editing trick that looks very cool to the viewer. I've only employed this technique once before, in a segment where I snapped my fingers and the lights came on. I'm excited to try this again and see how it's received.

We shoot the stand-up all the way through with nothing behind me, and then shoot it again with items in the background so the film editors will have both images to work with. The stand-ups go extremely well, and we're able to quickly move on to our interviews. I have five people scheduled today. Each of them is prepared and not one of them is nervous. What a relief! We fly through the interviews and I know it'll be an exceptional segment. I thank everyone for coming and then find a quiet corner in which to compose my script. On days like this I'm so glad to be a TV host, working with all these talented people.

Before long my scripting is finished and it's time to head downtown to my salon sponsor. I know cutting and coloring my hair will take several hours, so first I stop for a bite to eat. Most people don't think of visiting a salon as work, but it's definitely an important part of my job. Appearances and attractiveness are a big part of what television is all about, and TV people are expected to look great both on camera and off. Besides, you never know when you'll run into someone at the grocery store who recognizes you. A while back I interviewed with a major network for a prestigious hosting job. Six months later I bumped into the CEO and the vice president at a sushi restaurant. Fortunately I was dressed up for a shoot later that afternoon. Had I been wearing my sweats, I would have felt unprepared and unprofessional. After that chance meeting they contacted me several more times about potential job positions, so I'm sure the subsequent interest was due to my looking and acting like a professional in a social setting, not just during the job interview.

The salon is bustling with people, so I sit in the waiting area and pick up several entertainment magazines. It seems silly, but reading about pop culture helps me with my job. It gives me a springboard for my show and timely discussion topics with those I interview. Nearly everyone keeps up with pop culture—whether they admit it or not—and it's something that can be used to create conversation and calm nervous interviewees. When my stylist is ready, we head back to her chair and she begins to assess what I'll have done today. I've been coloring my hair a bit to highlight it and having it cut short for a more professional look. Visiting here every five weeks would cost me a fortune, so I'm grateful these folks agreed to be one of my sponsors. I often get a professional massage as well. I'm a firm believer in massage and acupuncture to relieve stress. The pace of being on television, always concerned with deadlines and ratings, can take a lot out of you. Setting aside time to relax is a valuable commodity in my business.

Once my hair is colored, my stylist trims off a few inches and I'm finished. It's exciting to be done earlier than usual, which means I can go home to unwind further. I plop down on the couch and flip on some entertainment news to catch up with that part of the business. I'll frequently watch interview-style shows as well, such as the one Oprah hosts, for ideas on opening my own show or ways to interview difficult people.

LESSONS/PROBLEMS

After what I was afraid would turn into a horrendously busy day, I made a vow to schedule some serious relaxation time. You have to do that regularly, or else risk going crazy.

WHEN FILMING A GREEN SCREEN SEQUENCE, IT TAKES TIME TO FIND THE RIGHT POSITION.

Day 21 **FEBRUARY 12**

PREDICTIONS

- Shoot a green-screen promo in the morning
- Have a business lunch with a producer regarding a new project
- Go on location for an afternoon shoot

DIARY

Today begins with a studio shoot that's scheduled to start at 9:00 A.M. I sit at my desk and fix my hair while awaiting the go-ahead. My managing producer announces that we'll be doing a green-screen stand-up with moving graphics added in behind me, but that's all I know. I imagine the graphic will contain information on show times, upcoming guests, spe-

GREEN SCREEN ALLOWS YOU TO HAVE GRAPHICS AND TEXT "FLY IN" TO THE FRAME WITH YOU, LIKE MAGIC.

cial promotions, and so on. We haven't done anything like this for a while, so it should be interesting. The key people involved in this shoot are tied up in meetings, so I kill time by scripting the interviews I did on Friday. This is an easy task, since everything went so smoothly. Once that's finished, I call over to the editing office to see if now is a good time to record my voice-overs for the show that was due on Friday.

I print out a copy of last week's script and head into the VO booth. My old office was set up so I could record everything myself, but since coming over here we need an editor to do the recording for us. I close the booth door behind me and put on my headset to await the prompt. After the signal I read my script with an authoritative voice and at a fairly

CHRIS HELPS ME GET MY HAIR SMOOTH SO THE GREEN BACKGROUND CANNOT BE SEEN THROUGH THE STRANDS.

speedy pace. We run through seven pieces in practically no time, although that's not always the case. Sometimes I read my voice-overs in too subdued a tone, which causes the editors to have me redo them. Voice-overs can add something special to a segment, or even save you when there aren't enough usable sound bites to make a nice package. I often throw them in when the interviewee is nervous or a less-than-adept public speaker. That allows me to say things for them the viewer needs to know.

As soon as this task is finished I hear my DP calling me to the studio for the promo shoot. The studio is fairly small, with its walls painted green so we can project any sort of graphic background behind the speaker in post-production. The number one rule when working with green screen is, "DO

NOT WEAR ANYTHING GREEN while on camera." It's important to stand out against the background color so it's easy for the film editor to "cut you out" and replace that blank background with whatever image they want the viewer to see in the finished shot. Today I'm wearing a cream-colored jacket and a gray skirt with black boots. We begin with me pretending to rest my arms on top of what will be the graphic that contains the information for my show. This proves tough to do without a reference point, so Chris takes a green cardboard square and asks me to lean against it. Now I have something physical that represents the graphical box to be plugged into the final image, which means I don't have to pretend as much. We do a few shots where I walk into the frame as if I'm dragging the graphic along behind me. We even try a few dancing motions where I prance around in front of the blank wall. It may seem silly, but major TV networks have produced these kinds of promos for years to capture the audience's attention.

Thirty minutes later we've wrapped up this shoot and I'm off to my lunch meeting. The DP on this new project, David, is already waiting for me. I worked with him a few months ago on an infomercial and found him to be highly skilled and very reliable, which puts me at ease regarding this upcoming gig. Fifteen minutes later the producer joins us and launches into the project details rather than starting off with small talk. He wants to know if I'm comfortable working with a TelePrompTer and an ear prompt, and also if I can deal with some potentially graphic content regarding murder. I say yes to all three questions while finding them rather unusual opening questions. Then he begins to relate the details behind the venture. Apparently a young homosexual college student was found dead at a popular gay bathhouse. The death was ruled a heart attack, despite clear signs of forced bodily harm. The organization that hired this producer to take on this project believes the victim was treated unfairly because of his minority racial status and sexual orientation. They're hoping this awareness campaign will cause the investigation to be re-opened as a murder case. I'm clearly intrigued to host this piece, especially since I don't get many offers to work on things like this. It sounds like a mix of the TV shows "Dateline" and "48 Hours." We discuss my day rate and I quote him what I believe is a fair number, something slightly under scale, at which point we agree to go forward together.

After lunch it's time to head back to the office for my afternoon assignment. The information I've been given on this shoot could fill an encyclopedia, which means the interviewees could prove very picky. I've read over the multiple pages of information, even questions they want me to ask. Well, at least that's less work I have to do. When we arrive at our destination, I'm delighted to discover these people to be exceedingly charming, especially since I was expecting iron-fisted field generals. We set up and knock out the interviews that include two resident families plus the director of the facility. The afternoon is over before I know it, having given me everything I need to put together a great show.

LESSONS/PROBLEMS

It's good to remember that the entertainment business includes plenty of hurry-up-and-wait moments. Shooting schedules aren't always written in stone, even though my green-screen shoot this morning had a fixed starting time. I have to be flexible, but also ready to go at a moment's notice when everyone else is set to proceed.

Day 22 *FEBRUARY 13*

PREDICTIONS
- Pick up more clothes for the week's shoots
- Script yesterday's shoot
- Participate in an afternoon shoot

DIARY

I have plenty of time to myself on days I pick up my new outfits, since the store doesn't open until 10:00 A.M. The manager who usually helps me has the day off today, so I pick out my own wardrobe this week. I search for things I'd buy if I could afford them, a fun chore. It's sometimes hard to imagine that I get to wear some of the most beautiful clothes I've ever seen. Once I pick out five complete outfits, I have just enough time to script yesterday's shoot at the office before meeting my friend Scott for lunch.

Scott and I are getting together to go over the details of my demo reel. He's created demos for me in the past, so he already understands the type of material that shows me in the best possible light. Today we're meeting to discuss formats and style options. We pore over the list of stand-ups that will be on the reel, deciding the order in which they should appear. Choosing the background music is pretty much all that remains to be done before he proceeds. I'm excited to have this new demo material appear on my Web site. We agree to reconvene later in the week to finalize everything.

When I return to the office, I run into the bathroom to change clothes for my afternoon shoot. Because we shoot three weeks in advance, I'm wearing a spring-type outfit even though it's freezing outside. This is occasionally inconvenient, and sometimes I forget about upcoming holidays since we shoot so far ahead of the calendar. Morris, my DP today, drives us to our afternoon location where we set up quickly for our interviews. I'm still feeling a bit under the weather and wish we could wrap as quickly as possible, but today's schedule calls for four interviews and two stand-ups. Hosting your own show means you don't get many days off, rarely have the luxury of a sick day, and minimal time to yourself. Still, there are plenty of trade-offs that make it a worthwhile vocation.

To speed things along I help Morris set up the audio, and soon our first subject is seated and ready to go. Most of the people I interview see the experience for exactly what it is—an interview. However, one gentleman doesn't seem to understand the concept. He acts as if the two of us are having a shoot-the-breeze conversation rather than something more formal, which causes a breakdown in structure. Relentlessly I try to steer him back to brief, concise answers, but it's obvious we're not going there today. Oh well, I can't win them all. My next two interviews prove more successful. Even though this is their first time on camera, these people clearly understand the purpose of keeping things short and to the point. When Morris starts to set up my stand-ups, we realize there's a problem with my outfit. My dress isn't exactly

A SLATE IS VITAL TO HAVING AN ORDERLY SHOOT.

risqué, but it's cut a bit lower than my usual fare. We try putting a sweater over it, shooting at a wider angle—anything to prevent the outfit from distracting viewers from what I'm saying. It starts to become pretty hilarious, but I know if we don't shoot these stand-ups today, doing so later in the week will interfere with the rest of our schedule. We do the best with what's available, even though we go through a few more takes than I'd like.

Back at the office, something is brewing again. Our managing producer is pacing back and forth in serious thought, and then he approaches Morris, Chris and me with a stressed-out look on his face. The boys upstairs—read: the decision makers who run the company—have decided we need to work weekends beginning this month. They feel we're missing too many events with our Monday-through-Friday shooting schedule. Our MP assures us it'll probably be one or two days a month at most, and in exchange we'll earn a half-day off during the week. I can barely believe what I'm hearing. It's bad enough to disrupt my weekends and interfere with my personal life, but getting only a half-day in exchange seems wrong. This announcement is met with cries of protest from all of us, but the bottom line is that if we don't comply, there are people out there who will. Sadly, we admit this is true. In the entertainment business, there are always people willing to do your job at a moment's notice. Of course, that doesn't mean there is someone of equal or better talent waiting to take your job, but nevertheless they're out there. I don't want to stop hosting my own show, but it's becoming harder to accept some of these new policies, nearly all of which have been negative.

LESSONS/PROBLEMS

Trying on outfits before appearing in them on camera is definitely a good idea. This will keep you from having embarrassing experiences. I'm troubled by having to work weekends, which will disrupt my personal life and also cut into the side gigs I want to do. I'm not sure if I can complain about it, but if I remain silent it could negatively affect my career.

Day 23 FEBRUARY 14

PREDICTIONS
- Do a shoot in the morning
- Audition during lunch to host Web broadcasts
- Script my morning interview

DIARY

It's very cold this morning, and the van is freezing as we drive to our early-morning shoot. Chris is my DP today, and we spend most of the ride listening to talk radio rather than discussing the recent announcement on having to work weekends. When we arrive at our location, our contact is not there and the office is closed. This is a bad sign, although honestly I forgot to make a reminder call last week to confirm the shoot. Many times I believe making these reminder calls is unnecessary. I mean, these people are all adults and they should keep track of the assignment on their own. After another fifteen minutes passes, I call my scheduling assistant. I learn she hadn't e-mailed them a reminder about the shoot either, which is potentially another nail in the coffin. We agree to wait twenty more minutes. Then if no one shows up, I'll call her back and we'll come up with Plan B.

It's inevitable that Chris and I start to talk about the weekend mandate, now that we have nothing else to do. He's as frustrated as I, and he's been with the show since its inception. Our company owns similar shows in other cities and we've heard those employees refused to comply, but it's too

I LIKE TO PRACTICE MY SUP BEFORE WE ROLL TAPE WHEN IT IS LONG OR HAS A FEW TECHNICAL FACTORS.

early to tell what might happen next. Just when we get into a drawn-out discussion, one of our interview subjects drives up. It turns out that no one knows anything about today's shoot, and no one else will be here except him. I call my office to give them the news, and they instruct me to get this interview in case we need the footage down the road. My DP gets the gear from the van and begins his setup so we can start as quickly as possible. I look at the clock and try not to worry about how long everything is taking, but I have a meeting in 90 minutes that I absolutely have to make. I'm auditioning for a host gig, the role of tour guide for several high-profile Web sites. The pay rate is excellent and all the reading will be done off TelePrompTer, which makes for easy work. I'm confidant in my ability to land this job, so it's imperative I wrap up quickly here.

MORRIS CHECKS OUT THE ANGLE TO MAKE SURE IT WILL LOOK APPEALING ON CAMERA.

Once we're set up and my interviewee is in place and slated, I begin with my questions. He starts off shakily and stumbles over a number of words, but I don't let that deter me. I know I have to bring out the best in him in order to get what I need and also finish on time. I smile at him warmly and use soft, soothing tones to convince him to relax. This can be one of the most tedious parts of hosting. It's not always easy or fun to get people to behave in a certain way within a narrow time frame. Now I know how Barbara Walters feels. Luckily my efforts work immediately, and I see his entire demeanor change. We speed through the rest of my questions and it becomes apparent we'll be finished in the nick of time. Normally I'd shoot a stand-up right away, but I let Chris know that we'll have to do it tomorrow from a generic location, since I have a very important appointment.

Conveniently my audition is only seven miles away, so I make it with time to spare. There's one person ahead of me, and I manage to overhear her performance. The script sounds complicated as she struggles through it. The man who's running the TelePrompTer appears to have it set rather fast, so I imagine speed-reading will be a pre-requisite for this gig. When it's my turn, I put on the microphone and stand on the mark in front of the camera, ready to go. They're shooting us against a green screen, so I'm guessing the casting director will superimpose a pleasant background before submitting these tapes to their client. We begin the audition as he rolls through the first of three short scripts, and I nail all the text cleanly. He seems pleased with my reading, so we move on to more difficult material. I have to admit, I stumble though one of them and misread the first line. He backs up the tape and I begin again. This time it goes smoothly and the delivery is exactly what they need. I feel good about taking the time to do this, and I'm certain that at some point it will lead to work.

I find time to grab a bite of lunch before returning to the office to script this morning's interview. Once the package is assembled and I've sent the script and the waiver over to our editors, I can finally relax a bit. I'm still stressed about getting my demo finished so it can be shopped around, so I call Scott and remind him of our mutually agreed-upon deadline. He comforts me by telling me everything is on schedule.

LESSONS/PROBLEMS

Having a personal life can be tough with a host's shooting schedule. Today I learned that keeping an appointment could prove stressful when people are not aware of a shoot or if they're not fully prepared. I need to come up with a Plan B for when things happen the way they did this morning. Whenever possible, I'll work hard to schedule my auditions and other meetings after standard work hours.

WE LIKE TO SET UP SHOTS SO THE VIEWER GETS A FEEL FOR THE THEME OR SEASON OF THE YEAR.

Day 24 FEBRUARY 15

PREDICTIONS

- *Finalize my demo reel*
- *Do both a morning and an afternoon shoot*
- *Return a phone call to a documentary producer*

DIARY

This has been such a busy week that I haven't had time to complete my demo reel. Scott informs me he's been spending a lot of time on it, and the editing portion of the task is nearly finished. We agree to meet over lunch to view it and make any necessary adjustments. I bring along several DVDs of recent work in case something needs to be replaced. When I arrive, the material is cued up and ready to go. As Scott begins to play the rough cut of the demo, I'm impressed with the job he's done. The images are color-corrected and the sound levels are evened out to reduce ambient noise. He took my audio-less New Year's Eve footage and created a fun, upbeat montage set to lively music. He also separated each type of segment with clever visual and musical transitions to alert the viewer that something different is about to play. One of the coolest things is the intro, which consists of four photos cribbed from my Web site that fade into the background as my printed name comes forward by way of a special effect. Overall it's both elaborate and beautifully done, the sort of fast-paced demo reel that's sure to capture all the right attention.

I'm nearly speechless. It's as if he's read my mind and knew exactly what I wanted. He included footage from my TV show plus material from the NHL All-Star weekend. I begin planning for its distribution, thinking I might be able to send it out as early as this weekend. After months of capturing all the footage, finding a time to put everything together and getting my press kit printed and shipped, having my demo completed is a huge weight off my shoulders. Now I can post the updated reel on my Web site as well.

After thanking Scott profusely, it's time to drive to my first shoot of the day. Today Morris is my DP and Zeif is our grip. Everything goes smoothly with the interviews, which makes me happy. We have the show open and close to shoot, however, and that proves to be more of a problem. The location is beautiful and my outfit works well with the décor, but the space we're working in is limited. We decide that I'll walk around a table that's covered with Spring Break-type items, since we're shooting a spring show. This type of shot isn't the most visually exciting, but for the sake of time we need to do something simple. We do three takes, and none of them is good. Then we do a fourth, a fifth and a sixth. My patience and my energy level are starting to wear thin. I stumble over my delivery and Morris has trouble with his camera movements. We can't seem to get on the same page at the same time, so we end up shooting the show open over and over and over. After our eleventh take, an almost unheard of number, we finally agree on a take we like and move on to shoot the close. This we nail in two takes, thank goodness.

As Morris shoots some interiors, I find a quiet place to return a call I missed earlier this morning. There's a message from a producer in Los Angeles to whom I'd previously submitted a demo reel. He's filming a documentary about bull riding and needs an experienced host. In his message he tells me my demo reel was the best of more than 150 submissions he'd received. That's good news, which makes me feel even better about my new demo reel. I can't wait for him to see that as well. He wants to fly out from Los Angeles and meet over lunch so we can discuss the project face to face, possibly within the next two or three weeks. I call back to let him know everything sounds great, but I only reach his voice mail. Therefore I e-mail him a short note to let him know I'm very interested in his project and would make myself available whenever he comes to town. I'm fairly sure this gig would cause me to use up all my vacation time, so I'm anxious to learn if a distributor or a network has already agreed to air the documentary. If so, spending my vacation days to shoot this program would be well worth it.

After Morris and Zeif finish their shooting, it's back in the van for another trip across town. It'll take us close to four hours to reach our next location, so I grab a nap while Morris drives. Our interviewees at this afternoon's shoot turn out to be friendly and highly prepared, which is great. Someone from their corporate office had briefed them on the kinds of questions I'd be asking, and they were also told what colors look best on camera. This makes the interview process so easy that I don't feel exhausted after two shoots and four hours of travel time. We film a quick stand-up. Even though it takes a few times to get it right—with people passing through our location while making too much ambient noise—at least it doesn't take 11 tries. After six takes we have a clean one with no outside noise and a great delivery. Tomorrow is a scheduled office day for me, which means no shoots, no interviews, and plenty of time to catch up on my paperwork.

LESSONS/PROBLEMS

Some days it can be harder than others to get in sync with your cameraperson, or to memorize your material if you're not using cue cards or a TelePrompTer. Today was frustrating for me, but that happens to everyone from time to time, so there's no need to panic.

Day 25 **FEBRUARY 16**

PREDICTIONS
- Script yesterday's shoots
- Log information into our computer system
- Call a few places regarding an upcoming theme show

DIARY

It's finally Friday, and it sure is nice to be working in the office. I have a lot of catching up to do, including scripting yesterday's shoots and entering the information into the computer system. It's a casual day as far as my wardrobe goes, so that makes things more comfortable. But I'm only at work a short time before noticing a strong air of negativity. Five minutes later my managing producer calls me into his office. This is rarely a good sign, especially when he shuts the door for privacy. It reminds me of being summoned to the principal's office when I was a kid. Apparently he's close to the end of his rope with the way things have gone since the layoffs, and he's decided to speak to each of us privately to clear the air and address our concerns. He thought my body language from our "working weekends" conversation the other day spoke volumes about my reluctance to jump on the bandwagon, and he wants to know my thoughts. I'm in a quandary. Answering honestly probably won't get me anywhere, but lying about it will just make me mad at myself later. I don't tell him how completely pointless I think it is, but I do say it's an inconvenience and I'm still unclear what role I'm supposed to play. I secretly wish I could write a ten-page report on how things could be run more efficiently and managed with greater consideration, but I know it would be ignored or else create conflict. My managing producer appears just as stressed about these changes, but I know he'd rather bend to every corporate whim rather than risk being replaced by someone else. It's clear he's doing his best to improve a negative situation, but he's also interested in knowing with whom my loyalties lie. We end on a moderately positive note, although I'm curious to find out how much notice I'll get when it's my turn to work a weekend.

After the meeting I sit down to script yesterday's interviews. Since my subjects were well prepared it makes things simpler, even though the sheer quantity of sound bites takes up a lot of my time. At the stroke of noon I stop in the middle of scripting my final package and head out to meet my sister Julie and my mother for lunch. It feels good to vent my frustrations regarding the way things are going at work, and I'm glad for their advice on how I should handle these issues. Just having their sympathy does wonders for my morale, and they both tell me to hang in there and focus on creating a stronger TV presence for myself. They know each gig is simply a springboard to the next one, until you finally reach the host job of your dreams. I leave lunch rejuvenated and ready to find something that will inspire me for the upcoming week.

In the afternoon it's time to start brainstorming new show-theme ideas. I'd originally thought about taking advantage of the Cirque de Soleil presence in town, but their day for media relations has come and gone. Consequently I'll need something else. The Body Worlds Exhibit is visiting our city following its tour of Europe. This is an artistic display of anatomical human specimens in plasticized form, including entire bodies as well as individual organs and transparent body cross-sections. I realize it may be a tough sell to my MP, since there is a tinge of death associated with the display— OK, so a lot of death. But this is a world-class exhibit, and it would be so incredibly interesting that I can see us capturing our viewers' attention in an instant. Instead of playing the "cool exhibit with educational value" card, I decide to go with the world-class exhibit angle. After all, only a few cities in this country will host it, and having things like that is a big media draw. I call up their marketing director and ask if they're interested in having us shoot there—hey, free publicity—and naturally they say yes. I catch hold of Chris while he's in the office to get his take on the subject. He's all for it, especially since it's something he wants to see for himself. He runs a few ideas by me about montages set to fun music and a few ways we can dress up the segment without making it seem morbid. Since he and I are both on board, all we need to do is convince our managing producer.

I overhear my MP in his office conversing cheerfully with another employee, so I take that as a good sign and forge ahead with my pitch. As soon as I mention the idea his face shows doubt. But as I continue talking about the prestige of the exhibit and the acclaim it brings to the city, I see a shift in his thinking. He gives me the green light to shoot it, even though he's apprehensive about what corporate management might think. Meanwhile, I'm confident we can make it a classy and interesting show.

LESSONS/PROBLEMS

Sometimes I have to come up with creative ways to pitch an idea that will get it approved by our corporate bosses, known collectively as the "suits." I'm successful more often than not, but it takes a whole lot more thought and planning than if I simply presented it in a predictable manner. As a host who's always looking for higher ratings and creative fulfillment, I know it's important to mix it up and do things that people might not expect.

Day 26 FEBRUARY 19

PREDICTIONS
- Make a trip to my clothing sponsor
- Have lunch with Scott to pick up my demos
- Do an afternoon shoot

DIARY

Arriving at my clothing sponsor's store, my usual helper isn't in yet so I begin picking out clothes on my own. I select two dresses and two tops with trousers, outfits that are designed for hotter weather. Because it's expected to be unseasonably warm this week, they'll fit right in. The office is really hopping when I arrive. Chris stops at my desk to discuss our afternoon shoot while I check my e-mail. We're headed for someplace new, a considerable drive from the office, so he suggests we leave earlier than usual to get back here at a reasonable hour. That means I'll need to go to lunch immediately, so I call Scott and ask him to move up our lunch date. Sometimes I have difficulty dealing with last-minute schedule changes. I know a host needs to be flexible, but it can be a tough pill to swallow when important things are going on in other areas of my life. Fortunately Scott is able to meet me an hour earlier than originally scheduled.

At lunch he hands me a bag of DVDs containing my new demo reel. He copied the ten discs I'd given him, and I'll probably need another 30 fairly soon. I'm relieved to have them in my possession; since they give me the freedom to

CHRIS LIKES TO SHOOT IN CREATIVE PLACES THAT CATCH VIEWER INTEREST.

crank up my press campaign and apply for other projects. I can't stress enough how important it is to have someone in your life familiar with your style and aware of your career aims. Scott has been that person for me, and he's been great with giving me advice and offering encouragement.

Then it's back to the office so I'll make the afternoon shoot on time. Even though I'm tempted to take a nap while Chris drives, instead we discuss what's been going on at the office. Our chat helps me gain perspective on how to deal with things, and consequently I'm in a better mood when we arrive on location. I miss a phone call while we're setting up, and after checking my messages I see it's the production company for whom I auditioned last Wednesday. They're giving me the gig, which is great news. Immediately I call them back and arrange for a shoot time. This proves a bit tricky, since I'm booked on eight shoots this week for my own show, but they're kind enough to carve out some time on Thursday evening to accommodate me.

Chris and I choose a spot for our interview and he starts setting up the lights and the camera. Meanwhile I concentrate on my interviewee, who's visibly nervous and admits as much. I try to get her mind off our interview, but she's having a hard time focusing on my words. This can be challenging, because oftentimes all a subject sees is a big black camera pointed at their face, plus someone like me drilling them with a million questions. She calms down a little once we start the Q & A session, except she stutters a few times and we're forced to backtrack for a cleaner sound bite.

Then we reset the scene for my stand-up. Chris likes to make creative and somewhat challenging camera movements, and sometimes he shoots the scene in a way that makes room for graphics and photos later on. Today we're doing a segment about security cameras and how people can use them in the home. I stand to one side of a framed shot, leaving plenty of room for our editors to insert graphics or a b-roll alongside me in postproduction. The segment works well, and I'm glad we decided to mix things up a bit. Our next scene involves a shot through a set of French doors, which I open in the middle of my spiel to "reveal" our subject matter—which will be added later electronically, of course. It takes us a few tries to get it right, but looking at the finished product we see it'll make a solid visual impression. Before heading back to the office, we cruise the area and capture some community footage. Once I've scripted today's interview, my workday will be finished.

LESSONS/PROBLEMS

It can be tricky to book outside gigs while continuing to work as a fulltime host, especially when I'm juggling a large shooting schedule. But it's always a good idea to try new things. Working on other projects keep me sharp, gives me proper perspective on my existing job, and just might open the door for something even better down the road.

Day 27 FEBRUARY 20

PREDICTIONS
- Two location shoots today, the first of which is a Mardi Gras theme party
- As today's second shoot involves children, it may prove difficult to get interviews

DIARY

I'm already expecting today to be a killer. Two shoots in one day are rough enough, but a Mardi Gras party plus a shoot with children is a recipe for exhaustion. We have several interviews scheduled at the party, including one with a V.I.P. who complained to my boss some time ago about not being filmed at an earlier event. As a result, we've made a special provision to get him on camera today. It's amazing how you have to beg some people to get in front of a camera, while others practically demand to have their face appear on television. There's nothing wrong with enjoying your own fifteen minutes of fame, but demanding it can come across as rude and self-centered.

We left the office so early it's apparent we've arrived at the party way too soon. None of the catering staff is here yet, which is not a good sign. We flag down a contact person and discover that yesterday the start time was pushed back two hours, not that anyone bothered to tell us. Morris suggests we do our stand-up now, before the guests arrive and chaos breaks out. I check my hair and makeup and we get ready for a tight close-up. While Morris and the grip are arranging the lights, I write a brief introduction—something basic about Mardi Gras and the beautiful cake that we'll shoot as soon as I'm finished. I blow our first take by stumbling over a word, but the second take is a keeper. We shoot a third simply because we have the time and it turns out to be the best one yet, so we decide to use it. I'm glad we have that out of the way before our interviewees show up.

I watch the catering staff file in as I sit down with my laptop and update my diary. I'm also expecting my managing producer to show up, so it'll look better if I'm typing away when he arrives. Meanwhile, Morris and the grip readjust the lights and check the audio, just to be sure everything's ready to go. As more people arrive I start to work the room, saying hello to everyone and seeking out our interviewees. Once I find one, I have him sign a waiver and get him seated in front of the camera. We generally shoot parties in the traditional "reporter" style, where I stand next to the camera and extend a handheld microphone to capture people's answers. There are times when we do a two-shot, with me standing next to my subject, but mostly we focus on the interviewee while I ask questions off-camera. Our first interview goes well. The moment it's over I see Mr. Demanding walk in the door. My managing producer is close behind him, reaching out to shake his hand. They both approach me, and I take a minute to explain our interview instructions: look at me rather than into the camera, answer in complete sentences, and sign the waiver that gives us consent to put people on television. He mentions he had a difficult time locating the party site, having found our directions faulty. UGH! It can be such a challenge to smile at someone when they find it difficult to return the favor. I set up his microphone and we begin our Q & A. He does quite well, which surprises me. At least the sound bites I'm capturing will be good ones.

Eight interviews later we're packed up and headed for lunch. We stop at a pizza place, and the restaurant's wait staff recognizes us from television. This doesn't happen all that often, but it's a nice feeling when it does. We load up on food from the buffet, eat at lightning speed, and then it's back on the road for our second shoot. We're covering a masterplanned community that boasts some amazing children's classes for its residents, including creative movement, ballet and tap. That last item makes me a bit nervous; I can already imagine the headache of hearing dozens of five-year-olds wearing tap shoes. We're early, so our contact person serves us drinks and snacks while my crewmates hang out and I return a few phone calls. When squeezing two shoots into a single day, it's often difficult to find time for those personal errands like checking e-mail, making phone calls, and running to the bank or the grocery store.

When classes begin we head upstairs to shoot some b-roll, which includes a scene of the instructor teaching the lesson while the children dance. Meanwhile I begin grabbing parents and asking them questions about the community, the classes, and the instructor. It proves harder to get them on camera than I'd originally hoped. Many parents prove camera-shy or aren't dressed appropriately for TV, since they didn't know in advance we'd be here. I don't get as many good sound bites as I'd like, and one of my best interviews becomes practically unusable. I begin to get a great sound bite from a parent, but her daughter starts picking her nose right in the middle of our interview. The footage will either have to be scrapped or covered with b-roll when it airs, which means we'll use only the audio portion. We move through class after class, and my potential interviewees dwindle away. I sit at my laptop, making notes and scripting the interviews. Some days this is a surprisingly unglamorous job.

LESSONS/PROBLEMS

It can be hard to get things organized in your life with two-a-day shoots. I try to schedule my personal errands and appointments around them, but it doesn't always work out the way I'd like. I also have to remember to be kind and gracious to even the most arrogant interviewees; it's a small world, after all.

GREEN SCREEN IS A FUN WAY TO SHOOT SHOW SUP'S AND COMMERCIALS, BUT CAN ALSO BE TEDIOUS WORK.

Day 28 FEBRUARY 21

PREDICTIONS

- Shoot a special promo for an awards dinner
- Spend a full day in the office to catch up on paperwork

DIARY

It's a relief to know I'll be at my desk all day, except for a couple hours of studio time to shoot a promotional video for a national awards dinner taking place next month. This gives me a chance to log all my interviews. I feel better about getting scripts to my editors early and working as far ahead of schedule as possible. It's also a lot easier to return personal e-mails and phone calls from work rather than while I'm on the road. I begin by scripting yesterday's remaining interviews and also refreshing a few. I enter into our computer system all the vital stats from this week's shoots: names of the people I met, contact information for all the locations we visited, and any special information that needs to run during the segment. This takes a while, but I'm able to squeeze in lunch and even a short nap before shooting the promo video in the afternoon.

Even though our studio is equipped with a TelePrompTer, the producer for this particular spot feels it's not worth taking the time to set it up. Someone delivers the script to my desk, and it contains huge clumps of text. I'm stunned we're not using a TelePrompTer for this project, considering the amount of memorization I'll need to do instead. After stewing over it for a couple of minutes, I shrug it off. After all, it'll be a chance to work on my memorization and remind everyone here that I'm a seasoned professional who can tackle anything thrown my way.

Morris sets up the lights in the studio and our producer takes her seat in the director's chair. We're shooting against a green screen again, although no one has mentioned what the background will be. The producer had cue cards made up, but we soon realize how awkward it will sound if I read them on camera. I take a deep breath and prepare myself mentally. I tell myself it's no big deal, since I memorize things every day. Also, this is a new producer with whom I've never worked, and I want her to realize I know what I'm doing. We start off with the first paragraph, a short one, and there are problems with the green screen filtering through my hair. I readjust my coiffure and begin again. Only after I get this first line correct does the producer inform me we'll be doing each paragraph three times today—once for a wide shot, once for a medium shot, and once for a close-up. Beads of sweat start to form on my forehead. "Did she just say three times?" I ask myself. I take another deep breath and start in on the second paragraph. I need three or four takes with each paragraph our first time through, but by the second round I pretty much have them memorized. I experience a quick flush of pride when we finally wrap, and I can't help but wonder how long this might have taken someone else. The producer tells me three thousand people will view this piece next month, which is great. I hope the finished product turns out well, and that all the perspiring I've done under these lights won't make it look like a midsummer day in Phoenix.

Back at my desk I start working on my theme ideas again. I ring up my contact at a local spa to see if we can set up a time to shoot. No one answers so I leave a friendly, upbeat message with the hope of receiving a speedy reply. I also mail out DVDs to people who have requested copies from past shows. It's a nice gesture to send examples of their work to the people who have appeared on camera with me. I depend heavily on the kindness of producers to send me the footage I need for my demo, so I try to remember to that for someone else.

As the day comes to a close, I head to the gym for a much-needed workout. With the office layoffs I haven't been as dedicated as I should be, mainly because the increased workload and frantic shooting schedule has sapped much of my excess energy. Still, I know I have to keep myself looking great. I decide to take a yoga class tonight to unwind and stretch out my muscles. I have two shoots scheduled for tomorrow and I may even shoot a show open and close, so I'll need to look as polished as possible. Television hosts do all sorts of silly things to look their best, such as tanning, but staying attractive and physically fit are high on our priority list. Before going to bed I dig out the clothing I'll need for another gig I'm hosting tomorrow evening. The organizers have requested that I dress brightly in business and business-casual attire, so I pull out every business suit I own along with khaki trousers and button-down shirts. I set my clothes next to the front door so I won't forget them in the morning.

LESSONS/PROBLEMS

I'm often asked to do things that would be easier if done differently, such as using a TelePrompTer for today's promo shoot. I know employing a TelePrompTer would have been faster, easier and more efficient, but the decision-makers were against it. I'm glad I chose to accept the challenge rather than complain or start an argument.

Day 29 FEBRUARY 22

PREDICTIONS
- Two shoots today, each on opposite sides of town
- New shoot as a Web site spokesperson

DIARY
We have both a cameraperson and a grip for our shoots today, but it will be a challenge to reach both our locations on time. We're supposed to be at our first spot by 9:30, but I wake up from my nap in the van to discover we're lost and already ten minutes late. I try to call our contact for directions, but the number I have for her is wrong. When I ring the office to speak to our scheduler, she's in a meeting. We break out our directory to look up the location, but it can't be found on the map because it's a newly developed area. I try to keep my blood pressure in check and remain calm, but being late is a huge no-no unless it's due to outrageously bad traffic. We pull into a gas station to ask directions, which turn out to be quite helpful. When we finally arrive, our contact is already in her office. I apologize profusely for our tardy arrival, but she doesn't seem bothered by it. That's a huge relief, but it doesn't change the fact that we're on a tight time schedule and already running thirty minutes behind.

The guys set up the lights and camera and I furiously write out a stand-up that can be shot simply and quickly. I know this is going to be one of those "under the gun" days, so

CHRIS SHOOTS QUICKLY UNDER PRESSURE.

I go for a lighthearted stand-up that won't need any extra work to make it stand out. Once everyone is ready, I head back to my contact's office to retrieve her. She's visibly nervous, having never been in front of a camera, so I brace myself for yet another challenging interview. Sure enough, she stumbles over her words and interrupts herself multiple times. After some coaching and verbal encouragement it's still clear to me that this won't work, so I decide to build most of the segment out of voice-over and b-roll. We nail the stand-up in three takes, which makes us feel better about our time constraints, and the crew heads outdoors to shoot exteriors. I remain inside and script the interview, even though it's a stretch to call it that.

After a quick lunch it's back in the van for the second part of our road trip. I wake up from my catnap just in time to keep us from getting lost again, and we actually end up at the location half an hour early. I take five minutes to write another stand-up while the guys get the lights situated, and then we shoot it in three takes. We reset the scene for our interviews and the first subject is in the chair at the assigned time. Today I'm interviewing three women and one man. The women are all comfortable in front of the camera, several having appeared on television before. Their answers give me exactly what I need. The man does a reasonably good job as well, although an hour before our interview he'd had Lasik surgery on his eyes and they appear rather glassy-looking through the viewfinder. But even after day surgery, people don't want to miss appearing on television. It gives them instant credibility and just a little bit of fame.

I've arranged to have my mother drive me to the office from our second location, since Morris has a lot more to do before his shooting day is over. There I pick up my car and go to my third shoot of the day, a side gig. This is a studio shoot with a green-screen background, like yesterday's promo, but today the script is on a TelePrompTer. Once I'm on my mark, we roll through the first section to make sure the TelePrompTer speed is correct. The gentleman who's running the device has it set way too fast, so he agrees to slow it down. However, once we start rolling he speeds it back up, which causes the two producers overseeing the shoot to complain that I'm reading too fast. We make some adjustments and try again, but it's still too fast and I feel myself becoming frustrated, especially with the hot lights beating down on me. Many studios shut off the air conditioning because the noise can interfere with the audio. My best solution seems to involve going with the flow to see how things play out. Eventually we arrive at a speed that satisfies everyone, and I simply fly through the script. Every so often the producers make a minor change to the text, but we're able to zip through the material rather quickly. We break for five minutes and the cameraman downloads the footage onto his laptop before bringing up my image and sampling it against different backgrounds. In one shot I'm seen standing in an airplane hangar, while in another I'm at the beach. It's amazing what they can do with green screens. For this production I'm playing the role of online tour guide, walking across different Web pages to explain various features on the site. Just before I leave they ask me to return and work on another project, which gives me great satisfaction for a job well done.

LESSONS/PROBLEMS

Even though it can be crazy when working on outside projects, it's worth the effort to stretch professionally. I have to remind myself to be patient with other people and let those in charge do their job at a pace with which they're comfortable.

KEEPING YOUR ENERGY UP IN FRONT OF THE CAMERA IS KEY TO ENGAGING ON-SCREEN PRESENCE.

Day 30 FEBRUARY 23

PREDICTIONS

- Go to a morning shoot with four interviews, plus the show's open and close
- Record voice-overs
- Begin mailing out my demo/press kit

DIARY

When I arrive at this morning's location, having driven there on my own, Chris has also just pulled up and is beginning to unload our equipment. I know I have a show open and close that will need to be shot today, and our schedule is very tight. I'm fresh out of ideas, so I dig around in my car for a book or a magazine, anything that might supply me with a fresh concept. All I can find is the latest issue of *InStyle* magazine plus a small devotional book with scriptural quotes. I thumb through them both, hoping something will jump out at me. I like to use literary quotes when I need to sound intelligent or desperately need to segue into a segment. It sounds silly, but I've used this technique quite often. Here are two of my favorite quotes:

"There is no shortage of good days. It is good lives that are hard to come by." (Annie Dillard)

"Hollywood is a place where they'll pay you a thousand dollars for a kiss and fifty cents for your soul." (Marilyn Monroe)

I come across a quote that states, "Life is not about the dreams we dream, but rather the choices we make that determine our steps." I decide to use it in my opening welcome to the show, and the rest of my writing flows much better once I've settled on a lead-in. My show close is just as easy, where I remind everyone to make choices that count and to tune in next week.

We set up indoors for our interviews, and three of my four subjects seem confident and prepared. The fourth, however, is overly excited about being on television for the first time. Actually, he can't stop talking about it. As I work to hook up my equipment to the camera and my laptop, he keeps interrupting with a thousand different questions. I finally have to tell him I can't deal with him at the moment, but I'll be glad to coach him through the interview process as soon as I have time. This seems to work, and we're finally ready to roll. My first three interviews go off without a hitch. The last one proves difficult, but I was already expecting that. This guy is so thrilled about his first time on camera that I can't get him to speak in complete sentences. Earlier he'd made cue cards for himself, which he asks me to hold while he reads them. The results are so stilted it comes off as totally unnatural, so we ditch the cue cards. He forgets to smile, he stumbles over his words, and then he starts to sweat profusely. Moments like these make me wonder how other hosts handle the pressure. I decide to keep the interview as simple as possible and basically just hit the highlights of what he knows.

Once the interviews are wrapped, Chris sets up the shot for the show open and close while I use my laptop to script the interviews we've just done. Then Chris calls me over to check our lighting and sound levels, after which we slate the open and give it a try. During the first take I miss my mark. On the second one he messes up his camera movement. By the third take we're operating like a well-oiled machine, and it turns out great. The best piece of advice I can give anyone doing stand-ups is to keep the energy level high and a smile on your face, so long as it's appropriate. Doing multiple takes to capture the correct lighting, audio and video can really sap your energy, and that soon becomes evident on camera. I try to keep my stomach full and my mind focused in order to maintain the energy I need to have a great sounding segment.

Back at the office I record my voice-overs. In this new office I finally have a booth where I can record them without needing a technician. It's almost second nature to me as I put on my headset, adjust the microphone stand to the correct height, and check the audio levels. We use a computer program that records and saves the voice-overs to a server so our editors can pull them up as needed. Then it's time to log on and begin my recording. The great thing about doing them without the help of an editor is that I can redo them as many times as I want without feeling I'm taking up too much of someone else's time. Once everything is recorded I e-mail the editors to let them know everything is posted on the server. After returning to my desk it's time to brainstorm ways of sending out my press kit. Creating them has proven to be an arduous process, but I can't deny I'm excited to begin shipping them out the door. With all the ups and downs these past six weeks have brought, especially with the layoffs and moving into a new office, I still feel a sense of joy when I see my show on the air or receive a friendly e-mail from someone who has learned something from our show. It may not be the most glamorous job all the time, and it's certainly not the least stressful, but it's one I do with great pride.

LESSONS/PROBLEMS

Becoming a working television host takes practice, hard work and, of course, a fair amount of luck. But if it's a dream you hold in your heart, it can be achieved. For me, having a backup plan is not an option. Appearing on television is everything I want to do professionally, and it can't be bought for more money in another field. I'm not always thrilled about interviewing nervous people or rushing to meet one deadline after another, but I love turning on the television and seeing my hard work displayed for the rest of the world to enjoy. I hope that more people have the opportunity to experience this profession, paving the way for television hosts to make a difference—or at least entertain!

GLOSSARY

B-Roll
The background video shot to accompany voice-over scripts. The four main categories include *exterior* (e.g., the outside of a building), *interior* (e.g., the contents of a room), *community* (e.g., an amenity such as a swimming pool), and *area* (e.g., the surrounding countryside).

Close
A stand-alone piece (i.e., not part of a package) at the end of a show that reminds viewers of future air times and other pertinent details.

Demo Reel
A collection of work, presented in audio or video format (VHS, CD, DVD, etc.), that contains examples of a performer's best material. Demo reels are often sequenced in time to music to create greater impact.

Director of Photography (DP)
The person primarily responsible for setting up shots and operating the camera during a session.

Green Screen (known in television as chroma key)
A term for the technique of using an evenly lit, monochromatic background for the purpose of replacing it with a different image or scene. The term also refers to the visual effect resulting from this technique as well as the colored screen itself. For example, TV weathercasts

are usually shot in front of a green screen, where maps are inserted electronically behind the talent.

Grip
The member of a film production crew who adjusts sets, lighting and props, and who often assists the camera operator.

Managing Producer
The person ultimately responsible for overseeing all aspects of a television production. The managing producer is responsible for hiring the creative team and assisting in day-to-day operations.

Name Super
The graphic that shows the name and title of the people being depicted on camera.

Open (part of Open & Close)
The stand-up introducing the show and the first package, e.g., "Welcome to the Wednesday edition of Hot On! I'm your host, Steve Smith. The community of..."

Slate
A board that contains the identifying details of a take of a film, held in front of the camera at the beginning and the end of each take. One of its functions is to synchronize picture and sound for later editing.

Slug
The graphic across the bottom of the screen, posted for a majority of the package, which contains information for

that segment.

SOT(s)
An abbreviation that denotes the sound bites contained in a package. The initials stand for Sound On Tape. It is usually pronounced as a single word, "sot."

Sponsors
Businesses that provide goods or services to the show, often in exchange for a promotional mention during the broadcast.

Stand-up
The host's on-camera introduction to a package.

Tease
Situated at the end of each show block, either voice-over or on-camera comments that, you guessed it, tease upcoming segments so viewers will stay tuned during the break.

Voice-over
Spoken dialogue, often recorded separately from the visual material, which is played over images being shown on screen.

OTHER CAREER DIARIES

Career Diary of an Animation Producer
Sue Riedl
ISBN: 1-58965-011-5

Career Diary of an Animation Studio Owner
Joseph Daniels
ISBN: 1-58965-010-7

Career Diary of a Caterer
Jennifer Heigl
ISBN: 1-58965-031-X

Career Diary of a Composer
Patrick Smith
ISBN: 1-58965-024-7

Career Diary of a Dental Hygienist
Nancy Aulie
ISBN: 1-58965-042-5

Career Diary of a Fashion Stylist
Kim Maxwell
ISBN: 1-58965-038-7

Career Diary of a Marketing Director
Christa Bahr
ISBN: 1-58965-045-X

Career Diary of a Movie Producer
Robin Hays
ISBN: 1-58965-041-7

Career Diary of a Newspaper Reporter
Hamil R. Harris
ISBN: 1-58965-033-6

Career Diary of a Personal Trainer
Bobby Hall
ISBN: 1-58965-039-5

Career Diary of a Publication Design Director
Leon Lawrence III
ISBN: 1-58965-030-1

Career Diary of a Publicist
Abby Lovett
ISBN: 1-58965-032-8

Career Diary of a Social Worker
Diana R. Hoover
ISBN: 1-58965-034-4

Career Diary of a Teacher
Carol Anderson
ISBN: 1-58965-035-2

Career Diary of a TV Production Manager
Craig Thornton
ISBN: 1-58965-015-8

Career Diary of a Veterinarian
Christine D. Calder
ISBN: 1-58965-043-3

Career Diary of a Vocalist
Nikki Rich
ISBN: 1-58965-037-9

Career Diary of a Web Designer
C. R. Bell
ISBN: 1-58965-022-0

GARTH GARDNER COMPANY

GGC publishing

Washington DC, USA · London, UK

Available at bookstore everywhere. For more information on these and other titles in Gardner's Career Diaries visit

www.GOGardner